Rings and Nearrings

Kostia Beidar fishing in Hsiao Liu-chiu

Rings and Nearrings

Proceedings of the
International Conference of Algebra

in Memory of Kostia Beidar

Tainan, Taiwan
March 6−12, 2005

Editors

Mikhail Chebotar
Yuen Fong
Wen-Fong Ke
Pjek-Hwee Lee

Walter de Gruyter · Berlin · New York

Editors

Mikhail Chebotar
Department of Mathematical Sciences
Kent State University
Kent, OH 44242
U.S.A.
e-mail: chebotar@math.kent.edu

Weng-Fong Ke
Department of Mathematics
National Cheng Kung University
Tainan 701
Taiwan
e-mail: wfke@mail-ncku.edu.tw

Yuen Fong
Department of Mathematics
National Cheng Kung University
Tainan 701
Taiwan
e-mail: fong@mail.ncku.edu.tw

Pjek-Hwee Lee
Department of Mathematics
National Taiwan University
Taipei 106
Taiwan
e-mail: phlee@math.ntu.edu.tw

Keywords: d-free set, quasi-polynomial, prime ring, boolean valued models, orthogonal completion, semiprime ring, nearring, fixed point free groups, strongly prime and semiprime modules, Hopf algebra, Lie derivations, Lie isomorphisms, separativity problem, regular rings, radicals, nil rings

Mathematics Subject Classification 2000: 16-xx; 16R50, 16N60, 16Y30, 16D40, 16D50, 16S50, 16W10, 16W25, 16E50, 16N20, 16N40

♾ Printed on acid-free paper which falls within the guidelines of the ANSI
to ensure permanence and durability.

Library of Congress Cataloging-in-Publication Data

International Conference on Algebra (2005 : T'ai-nan shih, Taiwan)
  Rings and nearrings : in memory of Kostia Beidar : proceedings of the International Conference on Algebra, Tainan, Taiwan, March 6–12, 2005 / edited by Chebotar, Mikhail ... [et al.].
    p. cm.
  ISBN 978-3-11-019952-9 (hardcover : alk. paper)
  1. Baidar, K. I. (Konstantin I.), 1951– 2. Rings (Algebra) – Congresses. 3. Near-rings – Congresses. I. Chebotar, Mikhail. II. Title.
  QA247.I558 2005
  512'.4–dc22
                                                              2007012794

ISBN 978-3-11-019952-9

Bibliographic information published by the Deutsche Nationalbibliothek

The Deutsche Nationalbibliothek lists this publication in the Deutsche Nationalbibliografie; detailed bibliographic data are available in the Internet at http://dnb.d-nb.de.

Preface

This volume is dedicated to the outstanding ring theorist of our time, Kostia Beidar, a distinguished professor of National Cheng Kung University, one of the leading universities in Asia. It consists of seven papers related to the various kind of research work of Kostia. Written by the leading experts of these areas the papers are not only in the aim of mathematical sense, but also emphasize the versatile applications to other fields of mathematics.

Most papers are based on the talks that were presented in the memorial conference which was held in March, 2005 in NCKU. The speakers were Tomoyuki Arakawa (Japan), Tatiana Bandman (Israel), Matej Brešar (Slovenia), Chen-Lian Chuang (Taiwan), Miguel Ferrero (Brazil), Antonio Giambruno (Italy), Koichiro Harada (USA), Shigeru Kobayashi (Japan), Ching-Hung Lam (Taiwan), Tsiu-Kwen Lee (Taiwan), Ying-Fen Lin (Taiwan), Christian Lomp (Portugal), Leonid Makar-Limanov (USA), Wallace S. Martindale, 3rd (USA), Edmund Puczyłowski (Poland), Peter Šemrl (Slovenia), Lance Small (USA), Richard Wiegandt (Hungary), Robert Wisbauer (Germany), Efim Zelmanov (USA). We would like to use this possibility to thank all the speakers for coming to this very special conference.

The editors are pleased to acknowledge support and financial assistance for the conference by National Science Council of R.O.C. and National Center for Theoretical Sciences.

Further, we cannot have this volume come into fruition without the help of many others. Here, acknowledgement should go to those who have encouraged and supported financially and spiritually through years. These include Professor Chiou-Shing Wang (Kao Yuan University), Mr. Yu-Lam Chu (the General Director of Mr. Mathematics Pig Publisher), Dr. Chuang Leo (the President of Cheng-Lin Investment Co., Ltd., and Cheng-Lin Education Co., Ltd.). Finally, special mention must be made to Professor Ka Wai Fong who has been supportive for the past decade.

January 2007

Mikhail Chebotar
Yuen Fong
Wen-Fong Ke
Pjek-Hwee Lee

Contents

Functional identities and d-free sets: fundamental contributions of Kostia Beidar

Matej Brešar

Dedicated to the memory of K. I. Beidar

1. Introduction

In a relatively short period 1994–2004 Kostia Beidar published, jointly with different coauthors, a number of papers [1]–[23] that, in one way or another, all base on *functional identities* (FI's): either by creating the general theory of FI's, or by dealing with applications of this theory to other mathematical areas. His impact on FI's was, and still is, unmeasurable. Some of his works, especially [1] (introducing "general" FI's) and two papers with Chebotar [14, 15] (introducing d-free sets), were path-breaking. They gave the foundations of the general theory of FI's.

Roughly speaking, functional identities are identical relations in rings that involve arbitrary ("unknown") functions together with arbitrary elements from a ring; the usual goal when treating an FI is to describe either the form of these functions or (when this is not possible) the structure of the ring admitting the FI in question. This topic was initiated in the early 90's by the present author, studying various special FI's in a series of papers. We refer to survey articles [32, 34] for details concerning this early period in the development of FI's. Let us just mention that many of these results, especially those on commuting maps [28, 29] and (using the present terminology) those on (generalized) functional identities of degree 2 [30, 31], were suggesting that something deeper is hidden behind all these. Some applications of the results on commuting maps, particularly those concerning Lie isomorphisms and derivations [29], made the problem of creating some "general theory" of FI's really attractive. For several years we were searching for a proper setting for such a theory.

Math. Subj. Class. (2000): 16R50.
Key words and phrases: functional identity, d-free set, $(t; d)$-free set, quasi-polynomial, prime ring, maximal left ring of quotients.

The first important break-through in this direction was made by Misha Chebotar who studied the so-called generalized functional identities and generalized our result from [31] from $n = 2$ to a general n [35] (the reason for the name "generalized functional identities" is that one can view these identities as generalizations of generalized polynomial identities, while functional identities generalize polynomial identities). Chebotar's paper was soon followed by Beidar's fundamental work [1] where in particular our result from [30] was generalized from $n = 2$ to a general n. But more importantly, the right concept was found; in our opinion, in [1] Beidar finally gave the answer to the question what should be the right setting for the general theory. And indeed such a theory was created in this setting soon afterwards – in the papers [14, 15] Beidar and Chebotar introduced and studied the so-called d-free sets which are now considered as the central concept in the theory of FI's. In our opinion these papers are astonishing also from the technical point of view.

The development of FI's has been always closely connected with applications. In particular, almost all abstract theory from [14, 15] was later used in the proofs of complete solutions of long–standing Herstein's conjectures [37] concerning Lie maps of associative rings.

Our aim in this paper is to present a survey of those results that are particularly important in the general theory of FI's. In Section 2 we will consider the theory of d-free sets, basically just surveying the papers [14, 15]. Section 3 is devoted to Beidar's fundamental theorem from [1] (this is also the only result in this paper which will be proved) and to related subsequent results establishing d-freeness of some concrete classes of sets.

We shall omit several important topics, including all applications of FI's. Although this makes this paper a kind of a torso, we believe that on the other hand it would be difficult to include everything relevant into one article and so it is better to select a particular area. We have tried to choose those results that we find beautiful and that in our opinion adequately represent the mathematical legacy of Kostia Beidar.

2. Beidar-Chebotar theory of d-free sets

As mentioned, in this section we will survey two papers [14, 15] by Beidar and Chebotar on d-free sets. In order to make the paper easily accessible we shall formulate only simplified versions of some results, and omit presenting the most involved results. Moreover, we shall not give any proofs. An interested reader should therefore consult the original papers.

The concept that we are about to introduce concerns arbitrary subsets of arbitrary rings, and is an elementary one in a sense that only the very basic knowledge of algebra is necessary to understand it (at least technically). However, the definition is somewhat complicated and before stating it we have to introduce various notations.

Let $\mathcal{Q}$ be a ring with unity, and let $\mathcal{R}$ be a nonempty subset of $\mathcal{Q}$. In the first concrete situation that was studied (and the one that appears in many applications), $\mathcal{R}$ was a ring and $\mathcal{Q}$ was its maximal ring of quotients (what explains the background for the choice of symbols $\mathcal{R}$ and $\mathcal{Q}$), but in the general theory that we shall now outline, $\mathcal{Q}$ can be just any ring with 1 and $\mathcal{R}$ can be its arbitrary nonempty subset. By $\mathcal{C}$ we denote the center of $\mathcal{Q}$.

Let m be a positive integer. For elements $x_i \in \mathcal{R}$, $i = 1, 2, \ldots, m$, we set

$$\bar{x}_m = (x_1, \ldots, x_m) \in \mathcal{R}^m,$$

$$\bar{x}_m^i = (x_1, \ldots, x_{i-1}, x_{i+1}, \ldots, x_m) \in \mathcal{R}^{m-1},$$

$$\bar{x}_m^{ij} = \bar{x}_m^{ji} = (x_1, \ldots, x_{i-1}, x_{i+1}, \ldots, x_{j-1}, x_{j+1} \ldots, x_m) \in \mathcal{R}^{m-2};$$

here $\mathcal{R}^k$ denotes the Cartesian product of k copies of $\mathcal{R}$. Let $\mathcal{I}, \mathcal{J}$ be subsets of $\{1, 2, \ldots, m\}$, and for each $i \in \mathcal{I}$ and $j \in \mathcal{J}$ let

$$E_i : \mathcal{R}^{m-1} \to \mathcal{Q} \quad \text{and} \quad F_j : \mathcal{R}^{m-1} \to \mathcal{Q}$$

be arbitrary functions. If $m = 1$, then we can regard E_i's and F_j's as elements in $\mathcal{Q}$.

The basic *functional identities*, that for the first time appeared in this generality in Beidar's paper [1] (the case $m = 2$ was considered earlier in [30]), are:

$$\sum_{i \in \mathcal{I}} E_i(\bar{x}_m^i) x_i + \sum_{j \in \mathcal{J}} x_j F_j(\bar{x}_m^j) = 0 \quad \text{for all } \bar{x}_m \in \mathcal{R}^m, \tag{1}$$

$$\sum_{i \in \mathcal{I}} E_i(\bar{x}_m^i) x_i + \sum_{j \in \mathcal{J}} x_j F_j(\bar{x}_m^j) \in \mathcal{C} \quad \text{for all } \bar{x}_m \in \mathcal{R}^m. \tag{2}$$

For example, if $m = 3$, $\mathcal{I} = \{1, 2\}$, and $\mathcal{J} = \{2, 3\}$, (1) can be rewritten as

$$E_1(x_2, x_3)x_1 + E_2(x_1, x_3)x_2 + x_2 F_2(x_1, x_3) + x_3 F_3(x_1, x_2) = 0 \tag{3}$$

for all $x_i \in \mathcal{R}$.

We remark that (1) trivially implies (2), so one should not understand that (1) and (2) are satisfied simultaneously by the same maps E_i and F_j; each of the two identities should be treated separately.

The usual goal when facing a certain FI is to describe the maps appearing in it, that is, we consider an FI as an equation with maps as unknowns. Let us present a natural possibility when (1), and hence automatically also (2), is fulfilled. Suppose there exist maps

$$p_{ij} : \mathcal{R}^{m-2} \to \mathcal{Q}, \ \ i \in \mathcal{I}, \ j \in \mathcal{J}, \ i \neq j,$$
$$\lambda_k : \mathcal{R}^{m-1} \to \mathcal{C}, \ \ k \in \mathcal{I} \cup \mathcal{J},$$

(for $m = 1$ one should understand this as that $p_{ij} = 0$ and λ_k is an element in $\mathcal{C}$) such that

$$E_i(\overline{x}_m^i) = \sum_{\substack{j \in \mathcal{J}, \\ j \neq i}} x_j p_{ij}(\overline{x}_m^{ij}) + \lambda_i(\overline{x}_m^i), \quad i \in \mathcal{I},$$

$$F_j(\overline{x}_m^j) = -\sum_{\substack{i \in \mathcal{I}, \\ i \neq j}} p_{ij}(\overline{x}_m^{ij}) x_i - \lambda_j(\overline{x}_m^j), \quad j \in \mathcal{J}, \tag{4}$$

$$\lambda_k = 0 \quad \text{if} \quad k \notin \mathcal{I} \cap \mathcal{J}.$$

A straightforward computation indeed shows that (4) implies (1), that is, (4) is a solution of the equation (1). We call (4) a *standard solution* of (1) (as well as of (2)). For example, a standard solution of (3) is

$$E_1(x_2, x_3) = x_2 p_{12}(x_3) + x_3 p_{13}(x_2),$$
$$E_2(x_1, x_3) = x_3 p_{23}(x_1) + \lambda_2(x_1, x_3),$$
$$F_2(x_1, x_3) = -p_{12}(x_3) x_1 - \lambda_2(x_1, x_3),$$
$$F_3(x_1, x_2) = -p_{13}(x_2) x_1 - p_{23}(x_1) x_2;$$

here, $p_{12}, p_{13}, p_{23} : \mathcal{R} \to \mathcal{Q}$ and $\lambda_2 : \mathcal{R}^2 \to \mathcal{C}$ are arbitrary maps.

The cases when one of the sets $\mathcal{I}$ and $\mathcal{J}$ is empty are not excluded. We shall follow the convention that the sum over $\emptyset$ is 0. Thus, if $\mathcal{J} = \emptyset$ (resp. $\mathcal{I} = \emptyset$), (1) reads as

$$\sum_{i \in \mathcal{I}} E_i(\overline{x}_m^i) x_i = 0 \quad \text{for all } \overline{x}_m \in \mathcal{R}^m, \tag{5}$$

$$\sum_{j \in \mathcal{J}} x_j F_j(\overline{x}_m^j) = 0 \quad \text{for all } \overline{x}_m \in \mathcal{R}^m. \tag{6}$$

Note that the standard solution of (5) is $E_i = 0$ for each i, and the standard solution of (6) is $F_j = 0$ for each j.

We are now in a position to introduce the basic notion of the theory of FI's.

Definition 2.1. A set $\mathcal{R}$ is said to be a *d-free subset of* $\mathcal{Q}$, where d is a positive integer, if the following two conditions hold for all $m \geq 1$ and all $\mathcal{I}, \mathcal{J} \subseteq \{1, 2, \ldots, m\}$:

(a) If $\max\{|\mathcal{I}|, |\mathcal{J}|\} \leq d$, then (1) implies (4).
(b) If $\max\{|\mathcal{I}|, |\mathcal{J}|\} \leq d - 1$, then (2) implies (4).

So, roughly speaking, on d-free sets the FI's (1) and (2) have only standard solutions provided that $|\mathcal{I}|$ and $|\mathcal{J}|$ are small enough, i.e. they do not exceed d (resp. $d - 1$). It follows easily from the definition that these standard solutions are *unique*.

Note that, in view of (a), (b) can be replaced by (b') If $\max\{|\mathcal{I}|, |\mathcal{J}|\} \leq d - 1$, then (2) implies (1). It turns out that the conditions (a) and (b) are often equivalent, but not in general [33].

A trivial remark: If $\mathcal{R}$ is d-free, then it is also d'-free for every $d' < d$. Usually we are interested in the largest d such that $\mathcal{R}$ is d-free. In the ideal situation $\mathcal{R}$ is d-free for every positive integer d.

A reader facing the definition of a d-free set for the first time might wonder whether d-free sets actually exist, and of course why to consider these complicated identities in the first place. One cannot answer the latter question immediately; let us just assure the reader that the notion of a d-free set has proved to be extremely useful. The first question has a clear answer: yes, d-free sets do exist, in fact there are plenty of them provided that $\mathcal{Q}$ satisfies certain conditions. All proofs of their existence that are known to us are based on Beidar's method that will be presented in the next section.

Of course such a definition does not come out of nothing. When Beidar and Chebotar introduced d-free sets in their paper [14] published in 2000, they knew that the concept of a d-free set is not an empty one, that is, they were aware of various concrete examples as well as of the applicability of the identities (1) and (2). But nevertheless the idea of studying *abstract* d-free sets appeared rather surprising at the time; later development has showed that it was a brilliant idea.

Concrete examples of d-free sets will be presented in the next section. Let us for a start point out some limitations with regard to d-freeness. Suppose that $\mathcal{R}$ is a nonzero commutative subset of $\mathcal{Q}$. Then $x_2 x_1 - x_1 x_2 = 0$ for all $x_1, x_2 \in \mathcal{R}$.

We can interpret this as that the FI

$$E_1(x_2)x_1 + E_2(x_1)x_2 = 0$$

has a nonzero, and hence a nonstandard solution, namely,

$$E_1(x_2) = x_2 \quad \text{and} \quad E_2(x_1) = -x_1.$$

Therefore $\mathcal{R}$ cannot be 2-free, and so also not d-free for every $d \geq 2$ (it may be 1-free, but this is not so interesting; the case $d = 2$ is the first nontrivial one). Accordingly, a commutative ring $\mathcal{Q}$ cannot contain 2-free subsets. Similarly one can show that if $\mathcal{Q}$ satisfies a polynomial identity of degree d, then $\mathcal{Q}$ does not contain d-free subsets. The question, however, whether $\mathcal{Q}$ contains $(d-1)$-free subsets, in particular whether $\mathcal{Q}$ itself is a $(d-1)$-free subset of itself, may be interesting.

One can view (multilinear) polynomial identities as very special examples of FI's of the type (1). Indeed, (1) reduces to a polynomial identity if all E_i's and F_j's are polynomials. But the theory of polynomial identities has rather different goals than the theory of functional identities, and as observations from the previous paragraph suggest, in PI rings one cannot hope for handling FI's easily. One could say, especially from the point of view of applications, that the two theories are complementary to each other, rather than that of FI's generalizes that of PI's.

Let us now present the basic results of the general theory of d-free sets. They can be, roughly speaking, divided into two groups: the results that yield new d-free sets from a given d-free set, and the results showing that on d-free sets one can handle more general FI's than those from the definition, i.e. (1) and (2).

We shall state just two sample results from the first group, both of them very useful and important. The first one has also a very simple statement. The proof, however, is nontrivial and is based on a generalization of the notion of a d-free set, which we shall avoid introducing in this expository article.

Theorem 2.2. *Let $\mathcal{P} \subseteq \mathcal{R} \subseteq \mathcal{Q}$ be nonempty sets. If $\mathcal{P}$ is a d-free subset of $\mathcal{Q}$, then $\mathcal{R}$ is also d-free.*

The next theorem also involves a concept more general than d-freeness. Although this concept is basically a technical one, perhaps not of great interest in its own right, we can not avoid it. It is too vital for the theory of FI's. This is the concept of a $(t; d)$-*free subset*. As it will be indicated in the next section, in order to show that a certain set is d-free, one is often forced to show more, namely that this set is $(t; d)$-free for some t. Let us introduce this concept.

We continue to assume that $\mathcal{R}$ is a nonempty subset of $\mathcal{Q}$, m is a positive integer, and $\mathcal{I}, \mathcal{J}$ are subsets of $\{1, 2, \ldots, m\}$. Now everything shall center round a fixed element $t \in \mathcal{Q}$. Let a, b be nonnegative integers, and $E_{iu} : \mathcal{R}^{m-1} \to \mathcal{Q}$, $i \in \mathcal{I}$, $0 \le u \le a$, and $F_{jv} : \mathcal{R}^{m-1} \to \mathcal{Q}$, $j \in \mathcal{J}$, $0 \le v \le b$, be arbitrary functions. The two identities that we shall now consider are clearly generalizations of (1) and (2):

$$\sum_{i \in \mathcal{I}} \sum_{u=0}^{a} E_{iu}(\bar{x}_m^i) x_i t^u + \sum_{j \in \mathcal{J}} \sum_{v=0}^{b} t^v x_j F_{jv}(\bar{x}_m^j) = 0 \quad \text{for all } \bar{x}_m \in \mathcal{R}^m, \qquad (7)$$

and

$$\sum_{i \in \mathcal{I}} \sum_{u=0}^{a} E_{iu}(\bar{x}_m^i) x_i t^u + \sum_{j \in \mathcal{J}} \sum_{v=0}^{b} t^v x_j F_{jv}(\bar{x}_m^j) \in \mathcal{C} \quad \text{for all } \bar{x}_m \in \mathcal{R}^m. \qquad (8)$$

Indeed, if $a = b = 0$ then we get (1) and (2). We define a *standard solution* of (7) and (8) as follows: there exist maps

$$p_{iujv} : \mathcal{R}^{m-2} \to \mathcal{Q}, \quad i \in \mathcal{I}, \; j \in \mathcal{J}, \; i \ne j, \; 0 \le u \le a, \; 0 \le v \le b,$$
$$\lambda_{kuv} : \mathcal{R}^{m-1} \to \mathcal{C}, \quad k \in \mathcal{I} \cup \mathcal{J}, 0 \le u \le a, \; 0 \le v \le b,$$

such that

$$E_{iu}(\bar{x}_m^i) = \sum_{\substack{j \in \mathcal{J}, \; v=0 \\ j \ne i}}^{b} t^v x_j p_{iujv}(\bar{x}_m^{ij}) + \sum_{v=0}^{b} \lambda_{iuv}(\bar{x}_m^i) t^v,$$

$$F_{jv}(\bar{x}_m^j) = - \sum_{\substack{i \in \mathcal{I}, \; u=0 \\ i \ne j}}^{a} p_{iujv}(\bar{x}_m^{ij}) x_i t^u - \sum_{u=0}^{a} \lambda_{juv}(\bar{x}_m^j) t^u, \qquad (9)$$

$$\lambda_{kuv} = 0 \quad \text{if} \quad k \notin \mathcal{I} \cap \mathcal{J}$$

for all $\bar{x}_m \in \mathcal{R}^m$, $i \in \mathcal{I}$, $j \in \mathcal{J}$, $0 \le u \le a$, $0 \le v \le b$.

Definition 2.3. A set $\mathcal{R}$ is said to be a $(t; d)$-*free subset of* $\mathcal{Q}$, where d is a positive integer and $t \in \mathcal{Q}$, if the following two conditions hold for all $m \ge 1$, all $\mathcal{I}, \mathcal{J} \subseteq \{1, 2, \ldots, m\}$, and all $a, b \ge 0$:

  (a) If $\max\{|\mathcal{I}| + a, |\mathcal{J}| + b\} \le d$, then (7) implies (9).
  (b) If $\max\{|\mathcal{I}| + a, |\mathcal{J}| + b\} \le d - 1$, then (8) implies (9).

Considering the case where $a = b = 0$ we see that if $\mathcal{R}$ is $(t; d)$-free for some $t \in \mathcal{Q}$, then it is also d-free. Although trivial, this observation is essential for us.

We say that $t \in Q$ is algebraic over C of degree $\leq n$ if there exist $c_0, c_1, \ldots, c_n$ $\in C$, not all zero, such that $c_0 + c_1 t + \ldots + c_n t^n = 0$. Now we are ready to state the second theorem.

Theorem 2.4. *Let $\mathcal{P}$ be a $(t; d+1)$-free subset of Q. Suppose that t is not algebraic over C of degree ≤ 2. Let $\epsilon \in \{1, -1\}$, and suppose that a set $\mathcal{R}$ is such that $tx + \epsilon xt \in \mathcal{R}$ for all $x \in \mathcal{P}$. Then $\mathcal{R}$ is a d-free subset of Q.*

This theorem is useful when one wants to establish the d-freeness of Lie or Jordan ideals of (Lie or Jordan) subrings of Q. Its proof is very complicated and rather long.

We now turn to the other group of results, showing that on d-free sets one can also settle other, more general and more complicated FI's than (1) and (2). Let us emphasize that the study of these identities was motivated by concrete problems like characterizing Lie homomorphisms etc.; most of the abstract theory of FI's, which we are now outlining, was developed because of applications that the authors had in their minds.

Consider the following situation. Let $\mathcal{S}$ be a nonempty set and let $\alpha : \mathcal{S} \to Q$ be an arbitrary (but fixed!) map. We shall write x^α for $\alpha(x)$. Let $m, \mathcal{I}, \mathcal{J}$ have the usual meaning, while the functions E_i and F_j now map from $\mathcal{S}^{m-1}$ into Q. We are now interested in FI's

$$\sum_{i \in \mathcal{I}} E_i(\bar{x}_m^i) x_i^\alpha + \sum_{j \in \mathcal{J}} x_j^\alpha F_j(\bar{x}_m^j) = 0 \quad \text{for all } \bar{x}_m \in \mathcal{S}^m, \tag{10}$$

$$\sum_{i \in \mathcal{I}} E_i(\bar{x}_m^i) x_i^\alpha + \sum_{j \in \mathcal{J}} x_j^\alpha F_j(\bar{x}_m^j) \in C \quad \text{for all } \bar{x}_m \in \mathcal{S}^m. \tag{11}$$

If $\mathcal{S} = \mathcal{R}$ and α is the identity map, then (10) and (11) coincide with (1) and (2), respectively. Now one can already guess how to define a *standard solution* of (10) and (11): there exist maps

$$p_{ij} : \mathcal{S}^{m-2} \to Q, \ i \in \mathcal{I}, j \in \mathcal{J}, i \neq j,$$
$$\lambda_k : \mathcal{S}^{m-1} \to C, \ k \in \mathcal{I} \cup \mathcal{J}$$

such that

$$E_i(\overline{x}_m^i) = \sum_{\substack{j \in \mathcal{J}, \\ j \neq i}} x_j^\alpha p_{ij}(\overline{x}_m^{ij}) + \lambda_i(\overline{x}_m^i), \quad i \in \mathcal{I},$$

$$F_j(\overline{x}_m^j) = -\sum_{\substack{i \in \mathcal{I}, \\ i \neq j}} p_{ij}(\overline{x}_m^{ij})x_i^\alpha - \lambda_j(\overline{x}_m^j), \quad j \in \mathcal{J}, \tag{12}$$

$$\lambda_k = 0 \quad \text{if} \quad k \notin \mathcal{I} \cap \mathcal{J}.$$

When do (10) and (11) have only standard solutions? The next theorem tells us that this depends *only* on the range of α.

Theorem 2.5. *Suppose that $\mathcal{S}^\alpha$ is a d-free subset of $\mathcal{Q}$.*

  (a) *If* $\max\{|\mathcal{I}|, |\mathcal{J}|\} \leq d$, *then* (10) *implies* (12).
  (b) *If* $\max\{|\mathcal{I}|, |\mathcal{J}|\} \leq d - 1$, *then* (11) *implies* (12).

Of course, if $\mathcal{S} = \mathcal{R}$ and α is the identity map, Theorem 2.5 is nothing but the restatement of the definition of a d-free set; it seems rather surprising that the two conditions (a) and (b) remain true for any map α.

The last topic we wish to discuss in this section is the theory of *quasi-polynomials*. We shall state only a sample result, and moreover only a special case of a more general theorem. There are two reasons for choosing this particular result: the first one is that it has turned out to be important because of applications, and the second one is that it has a simple and clear statement, which makes it possible for us to avoid introducing a complicated notation necessary for discussing more advanced results on quasi-polynomials.

Let us define a quasi-polynomial in an informal manner. As above, let $\alpha : \mathcal{S} \to \mathcal{Q}$ be a fixed map. We say that a map $P_1 : \mathcal{S} \to \mathcal{Q}$ is a quasi-polynomial of degree 1 if there exist $\lambda \in \mathcal{C}$ and $\mu : \mathcal{S} \to \mathcal{C}$, at least one of them nonzero, such that

$$P_1(x) = \lambda x^\alpha + \mu(x)$$

for all $x \in \mathcal{S}$. A quasi-polynomial of degree 2 is a map $P_2 : \mathcal{S}^2 \to \mathcal{Q}$ of the form

$$P_2(x_1, x_2) = \lambda_1 x_1^\alpha x_2^\alpha + \lambda_2 x_2^\alpha x_1^\alpha + \mu_1(x_2)x_1^\alpha + \mu_2(x_1)x_2^\alpha + \nu(x_1, x_2)$$

where $\lambda_1, \lambda_2 \in \mathcal{C}$, $\mu_1, \mu_2 : \mathcal{S} \to \mathcal{C}$, $\nu : \mathcal{S}^2 \to \mathcal{C}$, and at least one of them is nonzero. Now, a quasi-polynomial of degree 3 consists of summands such as

$$\lambda_1 x_1^\alpha x_2^\alpha x_3^\alpha, \quad \mu_1(x_1)x_2^\alpha x_3^\alpha, \quad \nu_1(x_1, x_2)x_3^\alpha, \quad \text{etc.}$$

with at least one "coefficient" (i.e. λ_1, μ_1, ν_1 etc.) nonzero. It should now be clear what we mean by a quasi-polynomial of degree n.

Theorem 2.6. *Let* $P : \mathcal{S}^{m-1} \to \mathcal{Q}$ *be a map. Let* $\lambda_i, \mu_i \in \mathcal{C}$ *be such that at least one of them is invertible, and define* $R : \mathcal{S}^m \to \mathcal{Q}$ *by*

$$R(\overline{x}_m) = \sum_{i=1}^{m} \lambda_i P(\overline{x}_m^i) x_i^\alpha + \mu_i x_i^\alpha P(\overline{x}_m^i).$$

Suppose that $\mathcal{S}^\alpha$ *is an* $(m+1)$*-free subset of* $\mathcal{Q}$. *If* R *is a quasi-polynomial, then* P *is a quasi-polynomial too.*

Of course P is of degree $m - 1$, unless all its coefficients are 0. We remark that the latter is equivalent to the condition that $P(\overline{x}_{m-1}) = 0$ for all $x_i \in \mathcal{S}$. This is a corollary to Theorem 2.5; in fact, to establish this it suffices to assume that $\mathcal{S}^\alpha$ is m-free (instead of $(m+1)$-free).

More general results in this area consider FI's involving summands such as

$$x_{i_1}^\alpha \dots x_{i_p}^\alpha P(x_{j_1}, \dots, x_{j_q}) x_{k_1}^\alpha \dots x_{k_r}^\alpha$$

where P is an arbitrary (unknown) function. We refer to [15] for these results, their proofs and all details.

3. Beidar's fundamental theorem and related results

Let us first fix the notation. By $\mathcal{A}$ we denote a prime ring, and by $\mathcal{Q}_{ml}$ we denote its maximal left ring of quotients (incidentally, "lcft" is chosen by chance, we could also work with the maximal right ring of quotients). The center $\mathcal{C}$ of $\mathcal{Q}_{ml}$ is a field called the extended centroid of $\mathcal{A}$. These notions are studied in detail in the book [24] by Beidar, Martindale and Mikhalev, to which we shall occasionally refer in the sequel.

Given $t \in \mathcal{Q}$, we denote by $\deg(t)$ the degree of algebraicity of t over $\mathcal{C}$ (if t is algebraic over $\mathcal{C}$) or ∞ (if it is not algebraic). We set $\deg(\mathcal{A}) = \sup\{\deg(t) \,|\, t \in \mathcal{A}\}$. It is known that the condition $\deg(\mathcal{A}) \leq n < \infty$ is equivalent to the condition that $\mathcal{A}$ satisfies the standard polynomial identity of degree $2n$, and in this case $\mathcal{A}$ can be embedded into a ring of $n \times n$ matrices over a field.

Theorem 3.1 (Beidar's fundamental theorem). *Let* $\mathcal{A}$ *be a prime ring and let d be a positive integer. If* $\deg(\mathcal{A}) \geq d$, *then* $\mathcal{A}$ *is a d-free subset of* $\mathcal{Q}_{ml}$.

This result was published in Beidar's seminal paper [1] in 1998. The original formulation was of course different because the notion of d-freeness was introduced only later. Also, in [1] the additional assumption that all maps E_i and F_j from (1) and (2) are multiadditive was used, but in the subsequent paper [22] it was noticed that this assumption is redundant.

It is worth mentioning that the condition $\deg(\mathcal{A}) \geq d$ is really necessary for the d-freeness of $\mathcal{A}$; moreover, it turns out that both conditions are actually equivalent. We shall concentrate, however, only on the more important direction stated in the theorem.

Our goal is to give the proof of Theorem 3.1. In fact, we will establish a more general result concerning $(t; d)$-freeness (Theorem 3.8) from which Theorem 3.1 follows. The main idea of the proof is the same as in Beidar's original paper, though some improvements noticed in subsequent works (particularly in [22]) will be used.

Let us first gather together all necessary tools needed for the proof. The first result is an improved version of a well-known lemma by Martindale [39, Theorem 2 (a)] that was observed in [38, Lemma 1.2] (in fact, the authors stated the lemma for the central closure of $\mathcal{A}$ playing the role of $\mathcal{Q}_{ml}$, but the lemma remains true in this slightly more general situation).

Lemma 3.2. *Let $a_i, b_i, c_j, d_j \in \mathcal{Q}_{ml}$ be such that $\sum_{i=1}^{m} a_i x b_i + \sum_{j=1}^{n} c_j x d_j = 0$ for all $x \in \mathcal{A}$. If $b_1, \ldots, b_m$ are linearly independent over C, then each a_i is a linear combination of $c_1, \ldots, c_n$ over C.*

If all c_i's are zero, then we get the classical situation [39] with the conclusion $a_i = 0$ for each i.

The next result is due to Erickson, Martindale and Osborn [36, Theorem 3.1] (see also [24, Theorem 2.3.3]).

Lemma 3.3. *If $a_1, a_2, \ldots, a_n \in \mathcal{A}$ are linearly independent over C, then there exists $c_l, d_l \in \mathcal{A}, l = 1, \ldots, m$, such that*

$$\sum_{l=1}^{m} c_l a_1 b_l \neq 0 \quad and \quad \sum_{l=1}^{m} c_l a_k d_l = 0, \ k = 2, \ldots, n.$$

Unlike the two lemmas above, the next one was found when investigating FI's. It plays a crucial role in the theory of FI's in prime rings. In particular it shows why maximal left rings of quotients come into play. Its first version appeared in [31, Lemma 5A], while the version which we shall now state is a special case of [1, Lemma 2.9].

Lemma 3.4. *Let $\phi : \mathcal{A} \to \mathcal{Q}_{ml}$ be an additive map. Suppose there exist a nonzero $c \in \mathcal{A}$ such that $\phi(ycx) = cy\phi(x)$ for all $x, y \in \mathcal{A}$. Then there exists a unique $q \in \mathcal{Q}_{ml}$ such that $\phi(x) = cxq$ for all $x \in \mathcal{A}$.*

Let us just mention that the main idea of the proof is to consider the map

$$\sum_i x_i c y_i \mapsto \sum_i x_i \phi(y_i)$$

from the ideal generated by c into $\mathcal{Q}_{ml}$. It is clearly a left $\mathcal{A}$-module homomorphism, so one can connect it with the basic properties of maximal left rings of quotients (see e.g. [24, Proposition 2.1.7]). There are, however, several technical details one is forced to consider in order to establish the lemma (an obvious one is to prove that the above map is well-defined). But we omit discussing them here.

We are now ready to tackle FI's. As above, m will be a fixed positive integer, $\mathcal{I}$ and $\mathcal{J}$ will be subsets of $\{1, 2, \ldots, m\}$, a, b will be nonnegative integers, and E_{iu}, F_{jv} will be arbitrary maps from $\mathcal{A}^{m-1}$ into $\mathcal{Q}_{ml}$. Further, t will be a fixed element in $\mathcal{A}$.

We shall abbreviate the notation a little bit. For maps $E : \mathcal{A}^{m-1} \to \mathcal{Q}_{ml}$ and $p : \mathcal{A}^{m-2} \to \mathcal{Q}_{ml}$ we shall write

$$E^i \text{ for } E(\overline{x}^i_m) \quad \text{and} \quad p^{ij} \text{ for } p(\overline{x}^{ij}_m).$$

Thus, for example, we will write (7) as

$$\sum_{i\in\mathcal{I}} \sum_{u=0}^{a} E^i_{iu} x_i t^u + \sum_{j\in\mathcal{J}} \sum_{v=0}^{b} t^v x_j F^j_{jv} = 0,$$

and the first identity in (9) as

$$E^i_{iu} = \sum_{\substack{j\in\mathcal{J}, \\ j\neq i}} \sum_{v=0}^{b} t^v x_j p^{ij}_{iujv} + \sum_{v=0}^{b} \lambda^i_{iuv} t^v.$$

Further, for $H : \mathcal{A}^m \to \mathcal{Q}_{ml}$ we will write

$$H(x_k t) \text{ for } H(x_1, \ldots, x_{k-1}, x_k t, x_{k+1}, \ldots, x_m)$$

and for $E : \mathcal{A}^{m-1} \to \mathcal{Q}_{ml}$ we will write

$$E^i(x_k t) \text{ for } E(x_1, \ldots, x_{i-1}, x_{i+1}, \ldots, x_{k-1}, x_k t, x_{k+1}, \ldots, x_m)$$

(provided that $i < k$; if $i > k$ then one must exchange the order appropriately.)

First we consider FI's

$$\sum_{i\in\mathcal{I}} \sum_{u=0}^{a} E^i_{iu} x_i t^u = 0 \quad \text{for all } \overline{x}_m \in \mathcal{A}^m \tag{13}$$

and

$$\sum_{j\in\mathcal{J}}\sum_{v=0}^{b} t^v x_j F_{jv}^j = 0 \quad \text{for all } \bar{x}_m \in \mathcal{A}^m. \tag{14}$$

Lemma 3.5. *If* $\deg(t) \geq |\mathcal{I}| + a$, *then (13) implies that each* $E_{iu} = 0$. *Similarly, if* $\deg(t) \geq |\mathcal{J}| + b$, *then (14) implies that each* $F_{jv} = 0$.

Proof. These two statements are left-right symmetric, so we prove only the first one. The proof is by induction on $|\mathcal{I}|$. If $|\mathcal{I}| = 1$, say $\mathcal{I} = \{1\}$, then (13) reduces to

$$\sum_{u=0}^{a} E_{1u}^1 x_1 t^u = 0 \quad \text{for every } x_1 \in \mathcal{A}.$$

Now, since $1, t, \ldots, t^a$ are independent over $\mathcal{C}$ by assumption, for any fixed $x_2, \ldots,$ x_m we may apply Lemma 3.2 to conclude that each $E_{1u} = 0$. So assume that $|\mathcal{I}| > 1$, say $1, 2 \in \mathcal{I}$. Set

$$H(\bar{x}_m) = \sum_{i\in\mathcal{I}}\sum_{u=0}^{a} E_{iu}^i x_i t^u = 0$$

and note that

$$0 = H(x_1 t) - H(\bar{x}_m)t$$

$$= \sum_{\substack{i\in\mathcal{I},\\ i\neq 1}} E_{i0}^i (x_1 t) x_i + \sum_{\substack{i\in\mathcal{I},\, u=1}}^{a} \{E_{iu}^i(x_1 t) - E_{i,u-1}^i\} x_i t^u - \sum_{\substack{i\in\mathcal{I},\\ i\neq 1}} E_{ia}^i x_i t^{a+1}.$$

We are in a position to apply the induction assumption. Consequently,

$$E_{ia} = 0, \quad E_{iu}^i(x_1 t) - E_{i,u-1}^i = 0 \quad \text{for all } i \neq 1, u = 0, 1, \ldots, a,$$

and hence each $E_{iu} = 0$ for $i \neq 1$. Similarly, each $E_{1u} = 0$ since we can replace the roles of x_1 and x_2 and repeat the argument. $\qquad\Box$

The next lemma is needed for handling the condition (b) of Definition 2.3. It considers a more general situation than we are actually interested in. It is quite typical that, when treating FI's, the proofs force us to modify the conditions that we wish to consider.

Lemma 3.6. *Suppose there exist maps* $\mu_w : \mathcal{A}^m \to \mathcal{C}, 0 \leq w \leq c$, *such that*

$$\sum_{i\in\mathcal{I}}\sum_{u=0}^{a} E_{iu}^i x_i t^u + \sum_{j\in\mathcal{J}}\sum_{v=0}^{b} t^v x_j F_{jv}^j = \sum_{w=0}^{c} \mu_w(\bar{x}_m) t^w$$

for all $\bar{x}_m \in \mathcal{A}^m$. If $\deg(t) > \max\{a + |\mathcal{I}|, c + |\mathcal{J}|\}$, then each $\mu_w = 0$.

Proof. By induction on $|\mathcal{J}|$. Suppose first that $|\mathcal{J}| = 0$. Then

$$\sum_{i \in \mathcal{I}} \sum_{u=0}^{a} E_{iu}^i x_i t^u$$

commutes with with t; commuting it with t we therefore arrive at

$$\sum_{i \in \mathcal{I}} \sum_{u=0}^{a+1} G_{iu}^i x_i t^u = 0$$

for all $\bar{x}_m \in \mathcal{A}^m$, where

$$\begin{cases} G_{i0}^i = -t E_{i0}^i, \\ G_{iu}^i = -t E_{iu}^i + E_{i,u-1}^i, \quad u = 1, \ldots, a, \\ G_{i,a+1}^i = E_{ia}^i \end{cases}$$

for each $i \in \mathcal{I}$. Since $\deg(t) \geq |\mathcal{I}| + (a + 1)$ by assumption, Lemma 3.5 tells us that each $G_{iu} = 0$, which in turn implies that each $E_{iu} = 0$. Therefore

$$\sum_{w=0}^{c} \mu_w(\bar{x}_m) t^w = 0.$$

Since $\deg(t) > c$ it follows that each $\mu_w = 0$.

Now let $|\mathcal{J}| \geq 1$, say $1 \in \mathcal{J}$. Set

$$H(\bar{x}_m) = \sum_{i \in \mathcal{I}} \sum_{u=0}^{a} E_{iu}^i x_i t^u + \sum_{j \in \mathcal{J}} \sum_{v=0}^{b} t^v x_j F_{jv}^j = \sum_{w=0}^{c} \mu_w(\bar{x}_m) t^w$$

and consider $H(tx_1) - tH(\bar{x}_m)$. Note that this yields

$$\sum_{i \in \mathcal{I}} \sum_{u=0}^{a} G_{iu}^i x_i t^u + \sum_{\substack{j \in \mathcal{J}, \ j \neq 1}} \sum_{v=0}^{b+1} t^v x_j H_{jv}^j$$

$$= \mu_0(tx_1) + \sum_{w=1}^{c} \{\mu_w(tx_1) - \mu_{w-1}(\bar{x}_m)\} t^w - \mu_c(\bar{x}_m) t^{c+1}$$

for some functions G_{iu}'s and H_{jv}'s. Since $c + 1 + |\mathcal{J} \setminus \{1\}| = c + |\mathcal{J}|$, we are in a position to apply the induction assumption. Hence it follows

$$\begin{cases} \mu_0(tx_1) = 0, \\ \mu_w(tx_1) - \mu_{w-1}(\overline{x}_m) = 0, \\ -\mu_c(\overline{x}_m) = 0 \end{cases}$$

for all $\overline{x}_m \in A^m$. This readily implies that each $\mu_w = 0$. $\quad\square$

The main step in the proof is handling the condition (a) of Definition 2.3. We do this in the next lemma.

Lemma 3.7. *If* $\deg(t) \geq \max\{|\mathcal{I}| + a, |\mathcal{J}| + b\}$, *then* (7) *implies* (9).

Proof. Suppose that all E_{iu}'s are of a standard form (9) (this includes the condition that $\lambda_{iuv} = 0$ whenever $i \notin \mathcal{J}$). Then we see from (7) that

$$\sum_{j \in \mathcal{J}} \sum_{v=0}^{b} t^v x_j \left[F_{jv}^j + \sum_{\substack{i \in \mathcal{I}, \ u=0 \\ i \neq j}}^{a} p_{iujv}^{ij} x_i t^u + \sum_{u=0}^{a} \lambda_{juv}^j t^u \right] = 0,$$

and so we infer from Lemma 3.5 that all F_{jv}'s are of the form (9) too. Similarly, if all F_{jv}'s are of the form (9), then the same is true for all E_{iu}'s.

The proof is by induction on $|\mathcal{I}| + |\mathcal{J}|$. Lemma 3.5 covers the cases where $|\mathcal{I}| = 0$ or $|\mathcal{J}| = 0$. Therefore the case when $|\mathcal{I}| = 1 = |\mathcal{J}|$ is the first one that has to be studied. We will consider separately two subcases: when $\mathcal{I} = \mathcal{J}$ and when $\mathcal{I} \neq \mathcal{J}$.

In the first subcase we may assume that $\mathcal{I} = \mathcal{J} = \{1\}$. Thus we have

$$\sum_{u=0}^{a} E_{1u}^1 x_1 t^u + \sum_{v=0}^{b} t^v x_1 F_{1v}^1 = 0.$$

We are now in a position to apply, for any fixed $x_2, x_3 \ldots, x_m$, Lemma 3.2. Hence we get

$$E_{1u}(x_2, \ldots, x_m) = \sum_{v=0}^{k} \lambda_{1uv}(x_2, \ldots, x_m) t^v$$

for some $\lambda_{1uv}(x_2, \ldots, x_m) \in \mathcal{C}$, $0 \leq u \leq a$, $0 \leq v \leq b$. Thus, E_{1u}'s are of the form (9), which forces that the same is true for F_{1v}'s.

In the second subcase we may assume with no loss of generality that $\mathcal{I} = \{2\}$ and $\mathcal{J} = \{1\}$. Thus,

$$K(\overline{x}_m) = \sum_{u=0}^{a} E_{2u}^2 x_2 t^u + \sum_{v=0}^{b} t^v x_1 F_{1v}^1 = 0. \tag{15}$$

We claim that the E_{2u}'s are additive in x_1. Indeed, replacing x_1 by $x_1' + x_1''$ in (15) we get

$$\sum_{u=0}^{a} \left(E_{2u}^2(x_1' + x_1'') - E_{2u}^2(x_1') - E_{2u}^2(x_1'') \right) x_2 t^u = 0$$

(we have simplified the notation by omitting writing $x_3, \ldots, x_m$). Since $\deg(t) > a$ by our assumption, it follows from Lemma 3.2 that

$$E_{2u}^2(x_1' + x_1'') = E_{2u}^2(x_1') + E_{2u}^2(x_1''),$$

proving our claim. Now fix $0 \le v \le b$. Since $\deg(t) > b$, that is $1, t, \ldots, t^b$ are independent over $\mathcal{C}$, by Lemma 3.3 there exist $c_l, d_l \in \mathcal{A}$, $l = 1, 2, \ldots, n$, such that $\sum_{l=1}^{n} c_l t^w d_l = 0$ if $w \ne v$ and $c = \sum_{l=1}^{n} c_l t^v d_l \ne 0$. Using (15) it follows that

$$0 = \sum_{l=1}^{n} c_l K(d_l x_1) = \sum_{u=0}^{a} H_{2u}^2 x_2 t^u + c x_1 F_{1v} \tag{16}$$

where $H_{2u}^2 = \sum_{l=1}^{n} c_l E_{2u}(d_l x_1, x_3, \ldots, x_m)$. Thus H_{2u}'s are additive in x_1. Substituting $y c x_1$ for x_1 in (16) we get

$$\sum_{u=0}^{a} H_{2u}(y c x_1, x_3, \ldots, x_m) x_2 t^u + c y c x_1 F_{1v}^1 = 0.$$

Since, on the other hand, $cy(c x_1 F_{1v}^1) = -\sum_{u=0}^{a} c y H_{2u}^2 x_2 t^u$ by (16), it follows that

$$\sum_{u=0}^{a} \left(H_{2u}(y c x_1, x_3, \ldots, x_m) - c y H_{2u}^2 \right) x_2 t^u = 0.$$

Lemma 3.2 implies that

$$H_{2u}(y c x_1, x_3, \ldots, x_m) = c y H_{2u}(x_1, x_3, \ldots, x_m)$$

for all $u = 0, 1, \ldots, a$ and $y, x_1, x_3, \ldots, x_m \in \mathcal{A}$. At this point Lemma 3.4 can be applied. It follows that there is a unique element $p_u(\overline{x}_m^{12}) \in \mathcal{Q}_{ml}$ such that

$$H_{2u}(\overline{x}_m^2) = c x_1 p_u(\overline{x}_m^{12}).$$

In view of (16) we now have

$$cx_1 \left(\sum_{u=0}^{a} p_u^{12} x_2 t^u + F_{1v}^1 \right) = 0,$$

and so, since $\mathcal{A}$ is prime,

$$F_{1v} = - \sum_{u=0}^{a} p_u^{12} x_2 t^u \quad \text{for all } 0 \le v \le b.$$

Thus all F_{1v}'s are of the form (9) and so the same holds true for all E_{2u}'s. The proof of the second subcase is complete.

We may now assume $|\mathcal{I}| + |\mathcal{J}| > 2$, and with no loss of generality that $|\mathcal{J}| \ge 2$ and that $1, 2 \in \mathcal{J}$. We have

$$H(\overline{x}_m) = \sum_{i \in \mathcal{I}} \sum_{u=0}^{a} E_{iu}^i x_i t^u + \sum_{j \in \mathcal{J}} \sum_{v=0}^{b} t^v x_j F_{jv}^j = 0.$$

Computing $H(tx_1) - tH(\overline{x}_m)$ it follows that

$$\sum_{i \in \mathcal{I}} \sum_{u=0}^{a} G_{iu}^i x_i t^u + \sum_{\substack{j \in \mathcal{J}, \\ j \ne 1}} x_j F_{j0}^j(tx_1)$$

$$+ \sum_{\substack{j \in \mathcal{J}, \\ j \ne 1}} \sum_{v=1}^{b} t^v x_j \{ F_{jv}^j(tx_1) - F_{j,v-1}^j \} - \sum_{\substack{j \in \mathcal{J}, \\ j \ne 1}} t^{b+1} x_j F_{jb}^j = 0$$

for some maps G_{iu}. Noting that $|\mathcal{J} \setminus \{1\}| + b + 1 = |\mathcal{J}| + b$ we see that the degree condition on t is fulfilled. By the induction assumption we get

$$F_{jb}^j = - \sum_{\substack{i \in \mathcal{I}, \\ i \ne j}} \sum_{u=0}^{a} q_{iuj\,b+1}^{ij} x_i t^u - \sum_{u=0}^{a} \mu_{ju\,b+1}^j t^u, \quad j \ne 1,$$

$$F_{jv}^j(tx_1) - F_{j,v-1}^j = - \sum_{\substack{i \in \mathcal{I}, \\ i \ne j}} \sum_{u=0}^{a} q_{iujv}^{ij} x_i t^u - \sum_{u=0}^{a} \mu_{juv}^j t^u,$$

$$j \ne 1, \quad v = 1, 2, \ldots, b,$$

$$\mu_{juv} = 0 \quad \text{if } j \notin \mathcal{I}.$$

Beginning with F_{jb} and proceeding recursively, we derive from these identities that

$$F^j_{jv} = -\sum_{\substack{i \in \mathcal{I},\ u=0 \\ i \neq j}}^{a} p^{ij}_{iujv} x_i t^u - \sum_{u=0}^{a} \lambda^j_{juv} t^u, \quad j \neq 1$$

for appropriate p_{iujv} and λ_{juv} with $\lambda_{juv} = 0$ if $j \notin \mathcal{I}$. In a similar fashion, by computing $H(tx_2) - tH(\bar{x}_m)$, we obtain

$$F^1_{1v} = -\sum_{\substack{i \in \mathcal{I},\ u=0 \\ i \neq 1}}^{a} p^{i1}_{iu1v} x_i t^u - \sum_{u=0}^{a} \lambda^1_{1uv} t^u,$$

$\lambda_{1uv} = 0$ if $1 \notin \mathcal{I}$, and so all F_{jv}'s are of standard forms. Consequently, the same is true for all E_{iu}'s.   □

Theorem 3.8. *Let $\mathcal{A}$ be a prime ring. If $t \in \mathcal{A}$ is such that $\deg(t) \geq d$, then $\mathcal{A}$ is a $(t; d)$-free subset of $\mathcal{Q}_{ml}$.*

Proof. Lemma 3.7 shows that the condition (a) is fulfilled. Using Lemma 3.6 for $c = 0$ we see that (8) implies (7) provided that $\max\{|\mathcal{I}| + a, |\mathcal{J}| + b\} \leq d - 1$. Therefore (b) now follows from (a).   □

Note that Theorem 3.8 implies Theorem 3.1. Indeed, if $\deg(\mathcal{A}) \geq d$, then $\mathcal{A}$ contains an element t with $\deg(t) \geq d$, hence $\mathcal{A}$ is $(t; d)$-free, and so also d-free.

Theorem 3.1 became a prototype of results that can be obtained. We will now state (without proofs) various other results, some of them considerably more general than this theorem. But in some sense they all depend on it, either they were derived from it or at least in their proofs similar techniques were used.

Already in [1] Beidar discussed basic FI's on Lie ideals of prime rings. Theorem 2.4 enables one to do this in a very elegant way. Using it Beidar and Chebotar proved in [14] the following result.

Theorem 3.9. *Let $\mathcal{A}$ be a prime ring and let $\mathcal{L}$ be a noncommutative Lie ideal of $\mathcal{A}$. If $\deg(\mathcal{A}) \geq d + 1$, then $\mathcal{L}$ is a d-free subset of $\mathcal{Q}_{ml}$.*

Let $\mathcal{A}$ be a ring with involution $*$. By $\mathcal{S}$ (resp. $\mathcal{K}$) we denote the set of its symmetric (resp. skew-symmetric) elements. In [22] Beidar and Martindale initiated the study of FI's of the form

$$\sum_{i \in \mathcal{I}} E_i(\bar{x}^i_m) x_i + \sum_{j \in \mathcal{J}} x_j F_j(\bar{x}^j_m) + \sum_{k \in \mathcal{K}} G_k(\bar{x}^k_m) x^*_k + \sum_{l \in \mathcal{L}} x^*_l H_l(\bar{x}^l_m) = 0,$$

which was later continued in [7]. Together with some results from [14] all these led to the following theorem.

Theorem 3.10. *Let A be a prime ring with involution such that $\mathrm{char}(A) \neq 2$. If $\deg(A) \geq 2d + 1$, then both S and K are d-free subsets of $\mathcal{Q}_{ml}$. Moreover, if $\deg(A) \geq 2d + 3$, then every noncentral Lie ideal of K is a d-free subset of $\mathcal{Q}_{ml}$.*

Especially the proof of the last statement is very complicated and long.

So far in this section we have considered only subsets of prime rings, and have discussed their d-freeness with respect to $\mathcal{Q}_{ml}$. For various reasons, in particular because of applications, this setting deserves a special attention. However, as seen in Section 2, one can consider the question of d-freeness in arbitrary rings. In [4] it was shown that appropriate modifications of arguments known in the prime ring setting enable one to obtain analogous results in different classes of rings. As we have just experienced, the notion of the degree of algebraicity over the extended centroid plays an important role when dealing with FI's in prime rings. There is no obvious analogue of this notion in other rings. Anyhow, [4] introduces two such notions which successfully replace the notion of $\deg(\,.\,)$ in certain classes of rings. They are called the *fractional degree* and the *strong degree*.

Making use of the fractional degree, together with the Beidar-Mikhalev theory of orthogonal completions [25, 26, 27] (see also [24]), Theorem 3.1 was generalized in [4] as follows.

Theorem 3.11. *Let A be a semiprime ring. If A does not contain nonzero ideals satisfying the standard polynomial identity of degree $2d - 2$, then A is a d-free subset of $\mathcal{Q}_{ml}$.*

Theorem 3.1 of course also covers simple rings. So we know when they are d-free — but only with respect to $\mathcal{Q}_{ml}$. Using the notion of the strong degree, which in simple unital rings coincides with $\deg(\,.\,)$, one can obtain a much stronger assertion:

Theorem 3.12. *Let A be a simple ring with unity. If $\deg(A) \geq d$, then A is a d-free subset of every ring $\mathcal{Q} \supseteq A$ with the same unity.*

A similar result can be obtained for the ring $M_d(B)$ of $d \times d$ matrices over any unital ring B. The proof is also based on the strong degree.

Theorem 3.13. *Let B be a ring with unity and let $A = M_d(B)$. Then A is a d-free subset of every ring $\mathcal{Q} \supseteq A$ with the same unity.*

References

[1] Beidar, K. I. *On functional identities and commuting additive mappings,* Comm. Algebra **26** (1998), 1819–1850.

[2] Beidar, K. I.; Brešar, M.; Chebotar, M. A. *Generalized functional identities with (anti-) automorphisms and derivations on prime rings, I,* J. Algebra **215** (1999), 644–665.

[3] Beidar, K. I.; Brešar, M.; Chebotar, M. A. *Functional identities on upper triangular matrix algebras,* J. Math. Sci. (New York) **102** (2000), 4557–4565.

[4] Beidar, K. I.; Brešar, M.; Chebotar, M. A. *Functional identities revised: the fractional and the strong degree,* Comm. Algebra **30** (2002), 935–969.

[5] Beidar, K. I.; Brešar, M.; Chebotar, M. A. *Functional identities with r-independent coefficients,* Comm. Algebra **30** (2002), 5725–5755.

[6] Beidar, K. I.; Brešar, M.; Chebotar, M. A.; Fong, Y. *Applying functional identities to some linear preserver problems,* Pacific J. Math. **204** (2002), 257–271.

[7] Beidar, K. I.; Brešar, M.; Chebotar, M. A.; Martindale, W. S., 3rd *On functional identities in prime rings with involution II,* Comm. Algebra **28** (2000), 3169–3183.

[8] Beidar, K. I.; Brešar, M.; Chebotar, M. A.; Martindale, W. S., 3rd. *On Herstein's Lie map conjectures, I,* Trans. Amer. Math. Soc. **353** (2001), 4235–4260.

[9] Beidar, K. I.; Brešar, M.; Chebotar, M. A.; Martindale, W. S., 3rd. *On Herstein's Lie map conjectures, II,* J. Algebra **238** (2001), 239–264.

[10] Beidar, K. I.; Brešar, M.; Chebotar, M. A.; Martindale, W. S., 3rd. *On Herstein's Lie map conjectures, III,* J. Algebra **249** (2002), 59–94.

[11] Beidar, K. I.; Brešar, M.; Chebotar, M. A.; Martindale, W. S., 3rd. *Polynomial preserving maps on certain Jordan algebras,* Israel J. Math. **141** (2004), 285–313.

[12] Beidar, K. I.; Chang, S.-C.; Chebotar, M. A.; Fong, Y. *On functional identities in left ideals of prime rings,* Comm. Algebra **28** (2000), 3041–3058.

[13] Beidar, K. I.; Chebotar, M. A. *On Lie-admissible algebras whose commutator Lie algebras are Lie subalgebras of prime associative algebras,* J. Algebra **233** (2000), 675–703.

[14] Beidar, K. I.; Chebotar, M. A. *On functional identities and d−free subsets of rings I,* Comm. Algebra **28** (2000), 3925–3951.

[15] Beidar, K. I.; Chebotar, M. A. *On functional identities and d−free subsets of rings II,* Comm. Algebra **28** (2000), 3953–3972.

[16] Beidar, K. I.; Chebotar, M. A. *On Lie derivations of Lie ideals of prime rings,* Israel J. Math. **123** (2001), 131–148.

[17] Beidar, K. I.; Chebotar, M. A. *On surjective Lie homomorphisms onto Lie ideals of prime rings,* Comm. Algebra **29** (2001), 4775–4793.

[18] Beidar, K. I.; Chebotar, M. A.; Mikhalev, A. V. *Functional identities in rings and their applications,* Russian Math. Surveys **59** (2004), 403–428.

[19] Beidar, K. I.; Fong, Y. *On additive isomorphisms of prime rings preserving polynomials,* J. Algebra **217** (1999), 650–667.

[20] Beidar, K. I.; Fong, Y.; Lee, P.-H.; Wong, T.-L. *On additive maps of prime rings satisfying Engel condition,* Comm. Algebra **25** (1997), 3889–3902.

[21] Beidar, K. I.; Lin, Y.-F. *On surjective linear maps preserving commutativity,* Proc. Roy. Soc. Edinburgh Sect. A **134** (2004), 1023–1040.

[22] Beidar, K. I.; Martindale, W. S., 3rd. *On functional identities in prime rings with involution,* J. Algebra **203** (1998), 491–532.

[23] Beidar, K. I.; Martindale, W. S., 3rd.; Mikhalev, A. V. *Lie isomorphisms in prime rings with involution,* J. Algebra **169** (1994), 304–327.

[24] Beidar, K. I.; Martindale, W. S., 3rd.; Mikhalev, A. V. Rings with generalized identities, Marcel Dekker, Inc., 1996.

[25] Beidar, K. I.; Mikhalev, A. V. *Orthogonal completeness and algebraic systems,* Russian Math. Surveys **40** (1985), 51–95.

[26] Beidar, K. I.; Mikhalev, A. V. *Homogeneous boundness almost everywhere for orthogonal complete algebraic systems,* Vestnik Kievskogo Universiteta, Ser. Mat. Mekh. **27** (1985), 15–17 (Ukrainian).

[27] Beidar, K. I.; Mikhalev, A. V. *The method of orthogonal completeness in structure theory of rings,* J. Math. Sci **73** (1995), 1–44.

[28] Brešar, M. *Centralizing mappings and derivations in prime rings,* J. Algebra **156** (1993), 385–394.

[29] Brešar, M. *Commuting traces of biadditive mappings, commutativity-preserving mappings and Lie mappings,* Trans. Amer. Math. Soc. **335** (1993), 525–546.

[30] Brešar, M. *On generalized biderivations and related maps,* J. Algebra **172** (1995), 764–786.

[31] Brešar, M. *Functional identities of degree two,* J. Algebra **172** (1995), 690–720.

[32] Brešar, M. *Functional identities: A survey,* Contemporary Math. **259** (2000), 93–109.

[33] Brešar, M. *On d-free rings,* Comm. Algebra **31** (2003), 2287–2309.

[34] Brešar, M. *Commuting maps: A survey,* Taiwanese J. Math. **8** (2004), 361–397.

[35] Chebotar, M. A. *On generalized functional identities in prime rings,* J. Algebra **202** (1998), 655–670.

[36] Erickson, T. S.; Martindale, W. S., 3rd.; Osborn, J. M. *Prime nonassociative algebras,* Pacific J. Math. **60** (1975), 49–63.

[37] Herstein, I. N. *Lie and Jordan structures in simple, associative rings,* Bull. Amer. Math. Soc. **67** (1961), 517–531.

[38] Lee, P.-H.; Lin, J.-S.; Wang, R.-J.; Wong, T.-L. *Commuting traces of multiadditive mappings,* J. Algebra **193** (1997), 709–723.

[39] Martindale, W. S., 3rd. *Prime rings satisfying a generalized polynomial identity,* J. Algebra **12** (1969), 576–584.

Matej Brešar, Department of Mathematics, University of Maribor, PEF, Koroška 160, SI-2000 Maribor, Slovenia

E-mail address: `bresar@uni-mb.si`

Boolean valued models and semiprime rings

Chen-Lian Chuang

Dedicated to the memory of Prof. K. I. Beidar.

Semiprime rings can be considered as Boolean product of primes rings, as discovered by Beidar and Mikhalëv [1]. Therefore, many results of prime rings can be extended to semiprime rings. One of the most systematic methods for such generalizations seems to be Beidar and Mikhalëv's theory of orthogonal completion. The theory consists of a through analysis of central idempotents of semiprime rings combined with a successful application of logical methods. Algebraic constructions, such as quotient structures or substructures, are clearly powerful tools to analyze algebraic structures. But logical constructions, such as filters and ultraproducts, are mainly designed to deal with logical complexity, which is caused by the logical operations (connectives and quantifiers) and which is usually represented by the complexity of sentences. Algebraic operations, essential to algebras, are usually treated as primitive function symbols in logic, not specially analyzed, so that the roles of logical operations can manifest themselves and so that logical complexity can be focused. Therefore, logical complexity does *not* always give information of algebraic structures. Beidar and Mikhalëv discovered a natural Boolean structure of semiprime rings and established the connection between the algebraic construction "quotient rings" and the logical construction "reduced products". Our aim here is to explain this theory with sufficient motivations. We also propose a generalization.

We will use Boolean valued models as our logical tool. Boolean valued models were first used by Paul J. Cohen in his famous independence proof of Axiom of Choice and Continuum Hypothesis. He formulated his proof in terms of partially ordered sets instead of Boolean algebras. The explicit formulation of Boolean valued models was done by R. M. Solovay [8]. The generalization from two-valued models to Boolean valued models seems rather routine mostly. So all the literature known to the author merely indicates how it can be done without actually developing it. Some of our references are hence about two-valued models. A standard source on this subject is [3]. Beidar and Mikhalëv also generalized their discoveries in semiprime rings to the theory of orthogonally complete algebraic systems. This can also be considered as a representation theory of Boolean valued models. We wish to work on this interesting topics somewhere else.

1. Boolean valued models

By a language, we mean a set $\mathcal{L}$ of constant symbols $c_0, c_1, \ldots$, function symbols $f_0, f_1, \ldots$ and predicate symbols $P_0, P_1, \ldots$. Each function symbol and also each predicate symbol are associated with an integer ≥ 0, called their arity. Constant symbols may be considered as 0-ary function symbols. Our logical symbols are: $\vee$ (*or*), $\wedge$ (*and*), $\neg$ (*not*), $\rightarrow$ (*if ..., then ...*), $\leftrightarrow$ (*...if and only if ...*), $\forall$ (*for all ...*) and $\exists$ (*there exists ...*). We also fix an infinite set of variables $x_1, x_2, \ldots$.

Definition 1.1. Terms in a language $\mathcal{L}$ are defined inductively as follows:

(1) Constant symbols and variables are terms.
(2) If $t_1, \ldots, t_n$ are terms and if f is an n-ary function symbol in $\mathcal{L}$, then the expression $f(t_1, \ldots, t_n)$ is also a term.
(3) All terms are so obtained.

Definition 1.2. Formulae in a language $\mathcal{L}$ are defined inductively as follows:

(1) If $t_1, \ldots, t_n$ are terms and P is an n-ary predicate symbol in $\mathcal{L}$, then $P(t_1, \ldots, t_n)$ is a formula. We call such formula atomic.
(2) If θ_1 and θ_2 are formulae, then $\neg\theta_1$, $\theta_1 \wedge \theta_2$, $\theta_1 \vee \theta_2$ are all formulae.
(3) If x is a variable and if θ is a formula, then $\exists x\theta$ and $\forall x\theta$ are formulae.
(4) All formulae of $\mathcal{L}$ are so obtained.

For notational convenience, we define $\theta_1 \rightarrow \theta_2$ as $(\neg\theta_1) \vee \theta_2$ and $\theta_1 \equiv \theta_2$ as $(\theta_1 \rightarrow \theta_2) \wedge (\theta_2 \rightarrow \theta_1)$.

Let B be a complete Boolean algebra. By an n-ary B-predicate on the set A, we mean a function from A^n to B. Let $\mathcal{L}$ be a language and A a set. Assume the following hold:

(1) To each n-ary function symbol $f \in \mathcal{L}$, there is assigned an n-ary function $\tilde{f} : A^n \rightarrow A$.
(2) To each n-ary predicate symbol $P \in \mathcal{L}$, there is assigned an n-ary B-predicate $\tilde{P} : A^n \rightarrow B$.

The set A together with these $\tilde{f}$, $\tilde{P}$ is called a B-valued model of $\mathcal{L}$ and we write $\mathcal{A} = \langle A; \tilde{f}, \tilde{P} \rangle_{f, P \in \mathcal{L}}$. We call A the ground set of the B-valued model $\mathcal{A}$. Assume that $\mathcal{A}$ is a B-valued model. For $e \in B$, we denote $1 - e$ by e^c. For $a_1, \ldots, a_n \in A$ and for each formula $\varphi(x_1, \ldots, x_n)$ in variables $x_1, \ldots, x_n$, we define inductively $\|\varphi(a_1, \ldots, a_n)\| \in B$ as follows:

(1) If P is an n-ary predicate, for $a_1, \ldots, a_n \in A$ we define
$$\|P(a_1, \ldots, a_n)\| \overset{\text{def.}}{=} \tilde{P}(a_1, \ldots, a_n).$$

(2) If $\varphi = \neg\theta$, we define $\|\varphi\| \overset{\text{def.}}{=} 1 - \|\theta\|$,

(3) If $\varphi = \theta_1 \wedge \theta_2$, we define $\|\varphi\| \overset{\text{def.}}{=} \|\theta_1\| \wedge \|\theta_2\|$.

(4) If $\varphi = \theta_1 \vee \theta_2$, we define $\|\varphi\| \overset{\text{def.}}{=} \|\theta_1\| \vee \|\theta_2\|$.

(5) If $\varphi = \exists x \theta(x)$, we define $\|\varphi\| \overset{\text{def.}}{=} \bigvee_{a \in A} \|\theta(a)\|$.

(6) If $\varphi = \forall x \theta(x)$, we define $\|\varphi\| \overset{\text{def.}}{=} \bigwedge_{a \in A} \|\theta(a)\|$.

Assume that $\mathcal{L}$ has the equality predicate $=$. Then we require the following axioms for the equality predicate $=$ hold.

(Axiom 1) $\|\forall x (x = x)\| = 1$.

(Axiom 2) $\|\forall x \forall y (x = y \equiv y = x)\| = 1$.

(Axiom 3) $\|\forall x \forall y \forall z ((x = y \wedge y = z) \rightarrow x = z)\| = 1$.

(Axiom 4) For any n-ary function symbol $f \in \mathcal{L}$,

$$\|\forall x_1 y_1 \cdots x_n y_n ((x_1 = y_1 \wedge \cdots \wedge x_n = y_n) \rightarrow$$
$$f(x_1, \ldots, x_n) = f(y_1, \ldots, y_n))\| = 1.$$

(Axiom 5) For any n-ary predicate symbol $P \in \mathcal{L}$,

$$\|\forall x_1 y_1 \cdots x_n y_n \Big((x_1 = y_1 \wedge \cdots \wedge x_n = y_n) \rightarrow$$
$$\big(P(x_1, \ldots, x_n) \equiv P(y_1, \ldots, y_n)\big)\Big)\| = 1.$$

We always assume that $\mathcal{L}$ has the equality predicate $=$. A simple induction proves the following. (See [6, pp. 75–77] for example.)

Lemma 1.3. *The following formula always has the Boolean value* 1 *for any formula* $\varphi(x_1, \ldots, x_n)$ *in* $\mathcal{L}$

$$\forall x_1 y_1 \cdots x_n y_n \Big((x_1 = y_1 \wedge \cdots \wedge x_n = y_n) \rightarrow$$
$$\big(\varphi(x_1, \ldots, x_n) \equiv \varphi(y_1, \ldots, y_n)\big)\Big).$$

A family $e_i \in B$, $i = 0, 1, \ldots$, is called *orthogonal* or *pairwise disjoint* or an *antichain* if $e_i e_j = 0$ for $i \neq j$.

Definition 1.4. (1) A Boolean valued model $\mathcal{A}$ is said to satisfy the mixing principle if for any orthogonal $e_i \in B$ and for any $a_i \in A$, there exists $a \in A$ such that $\|a = a_i\| \geq e_i$ for all i. (See [2, p. 24].)

(2) We call the B-valued model $\mathcal{A}$ *full* if for each formula of the form $\exists x \theta(x)$, there exists $a \in A$ such that $\|\exists x \theta(x)\| = \|\theta(a)\|$, or equivalently if for each formula of the form $\forall x \theta(x)$, there exists $a \in A$ such that $\|\forall x \theta(x)\| = \|\theta(a)\|$. (See [4, p. 160].)

A full model is also said to satisfy the maximum principle [7], because the set $\{\|\theta(a)\| : a \in A\}$ has the maximum. All the known fullness of Boolean valued models are established by the following

Lemma 1.5. *If a Boolean valued model satisfies the mixing principle then it is full.*

Proof. Given an existential sentence $\exists x \varphi(x)$, we have

$$\|\exists x \varphi(x)\| \overset{\text{def.}}{=} \bigvee_{a \in A} \|\varphi(a)\|.$$

Let us well-order A as $a_0 < a_1 < \cdots$. Define $c_0 \overset{\text{def.}}{=} \|\varphi(a_0)\|$ and inductively

$$e_i = \|\varphi(a_i)\| - \bigvee_{j < i} \|\varphi(a_j)\|.$$

We see easily that e_i's are orthogonal and $\bigvee_i e_i = \bigvee_i \|\varphi(a_i)\| \overset{\text{def.}}{=} \|\exists x \varphi(x)\|$. By the mixing principle, there exists $a \in A$ such that $\|a = a_i\| \geq e_i$ for all i. So

$$\|\varphi(a)\| \geq \|\varphi(a_i)\| \wedge \|a = a_i\| \geq e_i.$$

This is true for all i, so $\|\varphi(a)\| \geq \bigvee_i e_i = \|\exists x \varphi(x)\|$. Clearly, $\|\exists x \varphi(x)\| \geq \|\varphi(a)\|$. So $\|\exists x \varphi(x)\| = \|\varphi(a)\|$, as asserted. $\square$

If B happens to be the two-element Boolean algebra $\{0, 1\}$, then we call the two-valued model $\mathcal{A}$ a classical model or simply a model. An n-ary Boolean predicate $\overline{P}$ in a model is simply called a predicate and is usually considered as a subset of A^n with the interpretation that $(a_1, \ldots, a_n) \in \overline{P}$ if and only if $\|\overline{P}(a_1, \ldots, a_n)\| = 1$. In this case, we also say that φ is true if $\|\varphi\| = 1$ and that φ is false if $\|\varphi\| = 0$. One of great values of Boolean models is that they give rise to classical models in a very systematic way. For this purpose, we need the following

Definition 1.6. By a *filter* of a Boolean algebra B, we mean a subset F of B satisfying the following

(i) $1 \in F$ and $0 \notin F$.
(ii) If $e, f \in F$ then $e \wedge f \in F$.
(iii) If $e \in B$ satisfies $e \geq f$ for some $f \in F$, then $e \in F$.

A maximal filter is called an *ultrafilter*.

Let F be a filter of B. Let $\mathcal{A}$ be a B-valued model of $\mathcal{L}$ with the ground set A. We define the binary relation $\sim$ on A by

$$a \sim b \quad \text{if and only if} \quad \|a = b\| \in F \ \text{for } a, b \in A.$$

We check easily that $\sim$ is an equivalence relation on A. We set $a_F \overset{\text{def.}}{=} \{b \in A : a \sim b\}$, the $\sim$-equivalence class determined by $a \in A$. Let $A_F \overset{\text{def.}}{=} \{a_F : a \in A\}$. For each n-ary function symbol $f \in \mathcal{L}$, we define an n-ary function f_F on A_F by

$$f_F(a_{1,F}, \ldots, a_{n,F}) \overset{\text{def.}}{=} (\tilde{f}(a_1, \ldots, a_n))_F \ \text{for } a_1, \ldots, a_n \in A.$$

For each n-ary predicate symbol $P \in \mathcal{L}$, we define an n-ary predicate P_F on A_F by

$$P_F(a_{1,F}, \ldots, a_{n,F}) \quad \text{if and only if} \quad \tilde{P}(a_1, \ldots, a_n) \in F \ \text{for } a_i \in A.$$

We check easily that f_F and P_F are well defined. The set A_F together with f_F, P_F for $f, P \in \mathcal{L}$ forms a two-valued model for $\mathcal{L}$, which we denote by $\mathcal{A}_F$. We call $\mathcal{A}_F$ the *reduced* product of the B-valued model $\mathcal{A}$ modulo the filter F. If F is an ultrafilter then we call $\mathcal{A}_F$ the *ultraproduct* of $\mathcal{A}$ modulo F. The truth value (0 or 1) of a given sentence in a ultraproduct is nicely related to its Boolean value.

Theorem (Łôs 1954). *Assume that $\mathcal{A}$ is a full B-valued model, where B is a complete Boolean algebra. Let F be a ultrafilter of B. For any formula $\varphi(\vec{x}) \overset{\text{def.}}{=} \varphi(x_1, \ldots)$ and for any sequence $\vec{a} \overset{\text{def.}}{=} (a_1, \ldots)$ with $a_i \in \mathcal{A}$, $\|\varphi(\vec{a})\| \in F$ if and only if $\varphi(\vec{a}_F)$ holds in $\mathcal{A}_F$.*

Proof. Again, we suppress $\vec{x}$, $\vec{a}$ and $\vec{a}_F$ in formulae for brevity of notations. We proceed by induction on formulae. It suffices to deal with the only three logical symbols $\neg$, $\wedge$ and $\exists$, for they define $\vee$ and $\forall$.

(1) If φ is atomic, the assertion follows from the definition of $\mathcal{A}_F$.

(2) Let φ be $\neg\theta$. The assertion follows from the following equivalences:

$$\neg\theta \text{ holds in } \mathcal{A}_F \xLeftrightarrow{\text{def.}} \text{It is false that } \theta \text{ holds in } \mathcal{A}_F$$
$$\Longleftrightarrow \text{It is false that } \|\theta\| \in F$$
$$\Longleftrightarrow \|\theta\|^c \in F,$$

where the second equivalence follows by the induction hypothesis on θ, and where the third equivalence follows from the maximality of the ultrafilter F.

(3) Let φ be $\theta_1 \wedge \theta_2$. The assertion follows from the following equivalences:

$$\theta_1 \wedge \theta_2 \text{ holds in } \mathcal{A}_F \xLeftrightarrow{\text{def.}} \text{Both } \theta_1 \text{ and } \theta_2 \text{ hold in } \mathcal{A}_F$$
$$\Longleftrightarrow \text{Both } \|\theta_1\| \in F \text{ and } \|\theta_2\| \in F$$
$$\Longleftrightarrow \|\theta_1 \wedge \theta_2\| \xlongequal{\text{def.}} \|\theta_1\| \wedge \|\theta_2\| \in F,$$

where the second equivalence follows by the induction hypothesis on θ, and where the third equivalence follows from the closure under intersection for filters.

(4) Let φ be $\exists x \theta(x)$. If $\exists x \theta(x)$ holds in $\mathcal{A}_F$, then $\theta(c_F)$ holds in $\mathcal{A}_F$ for some $c \in \mathcal{A}$, which by induction hypothesis is equivalent to $\|\theta(c)\| \in F$. But $\|\exists x \theta(x)\| \xlongequal{\text{def.}} \bigvee_{a \in \mathcal{A}} \|\theta(a)\| \geq \|\theta(c)\|$. So $\|\theta(c)\| \in F$ implies $\|\exists x \theta(x)\| \in F$. On the other hand, by the fullness of the B-valued model $\mathcal{A}$, $\|\exists x \theta(x)\| = \|\theta(b)\|$ for some $b \in \mathcal{A}$. So $\|\exists x \theta(x)\| \in F$ implies $\|\theta(b)\| \in F$. By induction hypothesis, $\theta(b_F)$ holds in $\mathcal{A}_F$. So $\exists x \theta(x)$ holds in $\mathcal{A}_F$ with the witness b_F. This completes the proof.   □

But truth values in reduced products are very complicate and it is simple for special sentences in the following form.

Definition 1.7. The set of Horn formulae in the language $\mathcal{L}$ is defined inductively as follows:

(1) All atomic formulae and their negations are Horn.
(2) If A and $A_1, \ldots, A_n$ are all atomic, then the formulae

$$(A_1 \wedge \cdots \wedge A_n) \to A$$

   is Horn.
(3) If φ_1 and φ_2 are Horn, then the formula $\varphi_1 \wedge \varphi_2$ is Horn.
(4) If $\varphi(x)$ is Horn, then the formulae $\exists x \varphi(x)$ and $\forall x \varphi(x)$ are Horn.
(5) All Horn formulae are so obtained.

Theorem (Horn 1951). *Assume that $\mathcal{A}$ is a full B-valued model, where B is a complete Boolean algebra. Let $\varphi(\vec{x}) \overset{\text{def.}}{=} \varphi(x_1, \ldots)$ be a Horn formula and $\vec{a} \overset{\text{def.}}{=} (a_1, \ldots)$ a sequence with $a_i \in \mathcal{A}$. For any filter F of B, if $\|\varphi(\vec{a})\| \in F$, then $\varphi(\vec{a}_F)$ holds in $\mathcal{A}_F$.*

Proof. For brevity of notations, we suppress $\vec{x}, \vec{a}$ and $\vec{a}_F$ in formulae. We proceed by induction on Horn formulae according to the definition:

(1) Let φ be atomic. If $\|\varphi\| \in F$, then φ holds in $\mathcal{A}_F$ by the definition of $\mathcal{A}_F$. If $\|\neg\varphi\| = 1 - \|\varphi\| \in F$, then $\|\varphi\| \notin F$. So φ is *not* true in $\mathcal{A}_F$ by the definition of $\mathcal{A}_F$. Since $\mathcal{A}_F$ is a two-valued model, $\neg\varphi$ is true in $\mathcal{A}_F$, as asserted.

(2) Let $\varphi \overset{\text{def.}}{=} (A_1 \wedge \cdots \wedge A_n) \to A$, where $A_1, \ldots, A_n, A$ are atomic, be such that $\|\varphi\| \in F$. That is,

$$F \ni \|\varphi\| = \|(A_1 \wedge \cdots \wedge A_n) \to A\| = \|A_1\|^c \vee \cdots \vee \|A_n\|^c \vee \|A\|.$$

Assume that $A_1, \ldots, A_n$ hold in $\mathcal{A}_F$. By the definition of $\mathcal{A}_F$, $\|A_i\| \in F$ for $i = 1, 2, \ldots, n$. So

$$F \ni \|\varphi\| \wedge \|A_1\| \wedge \ldots \wedge \|A_n\| = \|A_1\| \wedge \ldots \wedge \|A_n\| \wedge \|A\|.$$

Since $\|A\| \geq \|A_1\| \wedge \ldots \wedge \|A_n\| \wedge \|A\|$, we have $\|A\| \in F$. So A holds in $\mathcal{A}_F$ by the definition of $\mathcal{A}_F$. Therefore, if $A_1, \ldots, A_n$ hold in $\mathcal{A}_F$ then so does A. This means that $\varphi \overset{\text{def.}}{=} (A_1 \wedge \cdots \wedge A_n) \to A$ holds in $\mathcal{A}_F$, as asserted.

(3) Let $\varphi = \theta_1 \wedge \theta_2$, where θ_1, θ_2 are both Horn. If $\|\varphi\| \overset{\text{def.}}{=} \|\theta_1\| \wedge \|\theta_2\| \in F$, then we have $\|\theta_1\|, \|\theta_2\| \in F$. By the induction hypothesis, the two Horn formulae θ_1, θ_2 both hold in $\mathcal{A}_F$. So the Horn formula $\varphi = \theta_1 \wedge \theta_2$ also holds in $\mathcal{A}_F$.

(4) Let φ be $\forall x \theta(x)$, where $\theta(x)$ is a Horn formula. Assume that $\|\forall x \theta(x)\|$ is in F. For any $b \in \mathcal{A}$, we have $\|\theta(b)\| \geq \bigwedge_{c \in \mathcal{A}} \theta(c) \overset{\text{def.}}{=} \|\forall x \theta(x)\| \in F$ and hence $\theta(b) \in F$. By induction hypothesis, $\theta(b_F)$ holds in $\mathcal{A}_F$. This is true for any $b \in \mathcal{A}$. So $\forall x \theta(x)$ holds in $\mathcal{A}_F$. Finally, let φ be $\exists x \theta(x)$, where $\theta(x)$ is a Horn formula. By the fullness of the B-valued model $\mathcal{A}$, $\|\exists x \theta(x)\| = \|\theta(b)\|$ for some $b \in \mathcal{A}$. So if $\|\exists x \theta(x)\| \in F$, then $\|\theta(b)\| \in F$ and, by induction hypothesis, $\theta(b_F)$ holds in $\mathcal{A}_F$. With the witness b_F, $\exists x \theta(x)$ holds in $\mathcal{A}_F$, as asserted. This completes the proof. $\square$

2. Application to semiprime rings

Throughout this section, R is a semiprime ring. Given an ideal I of R, let $I^* \overset{\text{def.}}{=} \{a \in R : aI = 0\}$. Obviously, $I \cap I^* = 0$ and $I + I^*$ is an essential ideal of R. Given a subset S of R, let $\langle S \rangle$ be the ideal generated by S. We call $\langle S \rangle^{**}$ the closure of S and denote it by $\bar{S}$. We call S closed if $\bar{S} = S$. Clearly, S is closed if and only if $S = J^*$ for some ideal J of R. So a closed set must be an ideal. Given a closed ideal I, we call I^* the complement of I. The pair of closed ideals I, I^* is called a complementary pair. For closed ideals I, J, define $I \vee J$ to be the closure of $I \cup J$. In this way, the family of closed ideals of R form a complemented lattice. Actually, it is a complete lattice: Given closed ideals I_i, $\bigvee_i I_i$ is merely the closure of $\bigcup_i I_i$.

Let $R_{\mathcal{F}}$ and Q be respectively the left and the symmetric Martindale quotient rings of R. The center C of $R_{\mathcal{F}}$ coincides with the center of Q and is called the extended centroid of R. Idempotents in C form a Boolean algebra B with respect to the operations $a \wedge b \overset{\text{def.}}{=} ab$, $a \vee b \overset{\text{def.}}{=} a + b - 2ab$. The extended centroid C is abundant in idempotents:

Definition 2.1. Given a subset S of R, let $I \overset{\text{def.}}{=} \langle S \rangle$, the two-sided ideal generated by S. Then $I \cap I^* = 0$ and $I + I^*$ forms an essential ideal of R. The map $\pi : a + b \mapsto a$ for $a \in I$ and $b \in I^*$ defines an idempotent $e \in C$ such that $ea = a$ for all $a \in S$. We denote this e by $\mathrm{E}[S]$.

The following gives the basic structure of elements in B.

Lemma 2.2. (1) $\mathrm{E}[S]$ is the minimal $e \in B$ such that $ea = a$ for $a \in S$.

(2) $\mathrm{E}[S] = \mathrm{E}[\bar{S}]$ and $\bar{S}$ is the maximal subset T of R such that $\mathrm{E}[T] = \mathrm{E}[S]$.

(3) Any $e \in B$ is of the form $\mathrm{E}[S]$ for some $S \subseteq R$.

(4) If $e \overset{\text{def.}}{=} \mathrm{E}[S]$, then $\bar{S} = \{r \in R : er = r\} = R \cap eR$, $\bar{S}^* = \{r \in R : er = 0\} = R \cap (1 - e)R$ and $\bar{S} \oplus \bar{S}^* = \{r \in R : er \in R\}$.

(5) $1 - \mathrm{E}[S] = \mathrm{E}[\langle S \rangle^*]$.

(6) $\mathrm{E}[S] \leq \mathrm{E}[T]$ iff $\bar{S} \subseteq \bar{T}$.

(7) $\bigvee_i \mathrm{E}[S_i] = \mathrm{E}[\bigcup_i S_i]$.

(8) $\bigwedge_i \mathrm{E}[S_i] = \mathrm{E}[\bigcap_i \bar{S}_i]$. For finitely many S_i, $\bigwedge_i \mathrm{E}[S_i] = \mathrm{E}[\prod_i \langle S_i \rangle]$.

Proof. (1) In the notation of the definition of $\mathrm{E}[S]$, write $e \overset{\text{def.}}{=} \mathrm{E}[S]$ for short. Suppose that $e' \in B$ also satisfies $e'a = a$ for $a \in S$. Since e' is central, $e'(a) = a$ for all $a \in I \overset{\text{def.}}{=} \langle S \rangle$. Write $r \in I + I^*$ as $r = a + b$, where $a \in I$ and $b \in I^*$. Then $e'e(r) = e'e(a) + e'e(b) = e'e(a) = e'(a) = a$. But we also have $e(a + b) = a$. So $e, e'e$, considered as right multiplications, coincide on the essential ideal $I + I^*$.

Then $e'e = e$ and hence $e' \geq e$. So e is the minimal element in B such that $er = r$ for $r \in S$.

(2) Let $I \overset{\text{def.}}{=} \langle S \rangle$. The map $\pi : a + b \mapsto a$ for $a \in I$ and $b \in I^*$ defines $E[S]$. The map $\pi' : a + b \in I^{**} + I^* \mapsto a$ for $a \in I^{**}$ and $b \in I^*$ defines the central idempotent $E[\bar{S}]$. But the maps π and π' coincide on the essential ideal $I + I^*$ of R and hence define the same element of $R_{\mathcal{F}}$. So $E[S] = E[I^{**}]$. Since $I^{**} = \bar{S}$, the first assertion follows. For the second assertion, let $T \subseteq R$ be such that $E[T] = E[S]$. Write $e \overset{\text{def.}}{=} E[S]$ for short. By the definition of $E[S]$, $er = 0$ for $r \in I^*$. But $er = r$ for $r \in T$. So $T \cap I^* = 0$. Then $TI^* \subseteq T \cap I^* = 0$ and hence $T \subseteq (I^*)^* = I^{**} = \bar{S}$, as asserted.

(3) and (4) Set $J \overset{\text{def.}}{=} \{r \in R : er \in R\}$. By definition of $R_{\mathcal{F}}$, J forms an essential ideal of R. Hence for $a \in R$, $aeJ = 0$ implies $ae = 0$. That is, $a \in J$ and $a = (1 - e)a$. So $(eJ)^* = (1 - e)J$ and analogously $((1 - e)J)^* = eJ$. Thus eJ and $(1 - e)J$ are complementary closed ideals of R. Clearly, $J = eI \oplus (1 - e)J$. For $a \in J$, $a = ea + (1 - e)a$. Via right multiplication, e maps $a = ea + (1 - e)a$ to ea. By definition, $e = E[eJ]$. Analogously, $E[(1 - e)J] = 1 - e$. Clearly, if $r \in eJ$ then $er = r$. Conversely, given $r \in R$, $er = r$ implies $r \in J$ and hence $r = er \in eJ$. So

$$eJ = \{r \in R : er = r\} = R \cap eR$$

and analogously,

$$(1 - e)J = \{r \in R : er = 0\} = R \cap (1 - e)R.$$

The assertion (3) follows by letting S be any subset of R satisfying $\bar{S} = eJ$, for then $e = E[\bar{S}] = E[S]$. Conversely, if $e = E[S]$ for a subset of R then $\bar{S} \supseteq eJ$ by the maximality in (2). On the other hand, since $e = E[S] = E[\bar{S}]$ by (2), we have

$$\bar{S} \subseteq \{r \in R : er = r\} = eJ.$$

So $\bar{S} = eJ$. Now, the rest of assertion (4) follows from what we have obtained above.

(5)–(7) are immediate consequences of (1)–(4) and Definition 2.1.

(8) Observe that

$$1 - \bigwedge_i E[S_i] = \bigvee_i (1 - E[S_i]) = \bigvee_i E[\langle S_i \rangle^*] = E[\bigcup_i \langle S_i \rangle^*].$$

So $\bigwedge_i \mathrm{E}[S_i] = 1 - \mathrm{E}[\bigcup_i \langle S_i \rangle^*] = \mathrm{E}[\langle \bigcup_i \langle S_i \rangle^* \rangle^*]$. The following equivalence holds for $a \in R$

$$a \in \langle \bigcup_i \langle S_i \rangle^* \rangle^* \quad \text{iff} \quad ar = 0 \text{ for } r \in \langle \bigcup_i \langle S_i \rangle^* \rangle$$

$$\text{iff} \quad ar = 0 \text{ for } r \in \sum_i \langle S_i \rangle^* \text{ since } \langle S_i \rangle^* \text{ are ideals}$$

$$\text{iff} \quad ar = 0 \text{ for } r \in \langle S_i \rangle^* \text{ and for all } i$$

$$\text{iff} \quad a \in \langle S_i \rangle^{**} \text{ for all } i \quad \text{iff} \quad a \in \bigcap_i \overline{S}_i.$$

So $\langle \bigcup_i \langle S_i \rangle^* \rangle^* = \bigcap_i \overline{S}_i$. The first equality follows as asserted.

For the second equality, we need a simple observation. For ideals I, J of R, if $IJ = 0$ then $I\overline{J} = 0$. *Reason*: We have $J \subseteq I^*$. Since I^* is closed and since $\overline{J}$ is the smallest closed ideal including J, we have $\overline{J} \subseteq I^*$, that is, $I\overline{J} = 0$ as observed. Now, set $I \overset{\text{def.}}{=} (\prod_i \langle S_i \rangle)^*$. So $IS_1 \cdots S_n = 0$. Applying the observation repeatedly, we see that $I\overline{S}_1 \cdots \overline{S}_n = 0$. But $\bigcap_{i=1}^n \overline{S}_i \subseteq \overline{S}_j$ for all $1 \le j \le n$. So $(\bigcap_{i=1}^n \overline{S}_i)^n \subseteq \overline{S}_1 \cdots \overline{S}_n$. We hence have

$$I(\bigcap_{i=1}^n \overline{S}_i)^n \subseteq I\overline{S}_1 \cdots \overline{S}_n = 0.$$

So $(I\bigcap_{i=1}^n \overline{S}_i)^n \subseteq I(\bigcap_{i=1}^n \overline{S}_i)^n = 0$. As R is semiprime, we get $I\bigcap_{i=1}^n \overline{S}_i = 0$. So $(\bigcap_{i=1}^n \overline{S}_i)^* \supseteq I$. On the other hand, since $\prod_i \langle S_i \rangle \subseteq \overline{S}_j$ for all $1 \le j \le n$, we have $\prod_i \langle S_i \rangle \subseteq \bigcap_{i=1}^n \overline{S}_i$. So $(\bigcap_{i=1}^n \overline{S}_i)^* \subseteq (\prod_i \langle S_i \rangle)^* \overset{\text{def.}}{=} I$. So we have

$$I = (\prod_i \langle S_i \rangle)^* = (\bigcap_{i=1}^n \overline{S}_i)^*.$$

This shows that the two ideals $\prod_i \langle S_i \rangle$ and $\bigcap_{i=1}^n \overline{S}_i$ have the same closure. (Actually, we have also shown that $\bigcap_i \overline{S}_i$ is closed in the first paragraph.) By (2) of this lemma, $\mathrm{E}[\prod_i \langle S_i \rangle] = \mathrm{E}[\bigcap_{i=1}^n \overline{S}_i]$. The second equality now follows from the first equality. $\square$

Note that (5)–(8) together show that the lattice B is isomorphic to the lattice of closed ideals of R. Also, by (7) or (8), B forms a complete Boolean algebra. We will make our ring R a Boolean valued model over B. We start with the equality predicate "=". For $a, b \in R$, we define

$$\|a = b\| \overset{\text{def.}}{=} \bigvee \{e \in B : ea = eb\}.$$

We have the following equivalence for $e \in B$:

$$ea = eb \text{ iff } e(a - b) = 0 \text{ iff } (1 - e)(a - b) = a - b \text{ iff } 1 - e \leq \mathrm{E}[a - b].$$

So $\|a = b\| = 1 - \mathrm{E}[a - b]$ and hence $\|a = b\|a = \|a = b\|b$. Therefore, we have

(2.1) $\|a = b\| \overset{\text{def.}}{=}$ the greatest $e \in B$ such that $ea = eb$.

So, if $\|a = b\| = 1$ then $a = \|a = b\|a = \|a = b\|b = b$. Clearly, the Boolean value thus assigned to the equality symbol "=" satisfies Axioms 1–3 in §1. In order to make our model full by Lemma 1.5, we want our model satisfies the mixing principle. We are thus led to the following

Definition 2.3. A subset S of the semiprime ring R is called orthogonally complete if for any maximal orthogonal $e_i \in B$ and for any $a_i \in S$, there exists $a \in R$ such that $e_i a = e_i a_i$ for all i. The element a is uniquely determined by e_i, a_i and will be denoted by $\sum^{\perp} e_i a_i$.

Clearly, the binary Boolean predicate "=" restricted to a subset S of R satisfies the mixing principle if and only if the subset S is orthogonally complete. It is shown in [1] that $R_{\mathcal{F}}$, Q and C are all orthogonally complete. We will hence assume that our ring R is orthogonally complete from now on. By the orthogonal completeness of R, if $e_i \in B$ form a maximal orthogonal set of B then $R = \prod_i e_i R$. If P is a prime ideal of R then P must include all $e_i R$ but one. This gives a simple description of minimal prime ideals of R.

Lemma 2.4. *Let R be an orthogonally complete semiprime ring. Let F be a filter of B. Set $F' \overset{\text{def.}}{=} \{1 - e : e \in F\}$.*

(1) *The set RF', consisting of finite sums $\sum_i r_i e_i$ for $r_i \in R$ and $e_i \in F'$, forms an ideal of R. We call RF' a filter ideal. For $r \in R$, we have $r \in RF'$ iff $\mathrm{E}[r] \in F'$.*

(2) *If F is an ultrafilter of B then RF' is a minimal prime ideal of R. Conversely, any minimal prime ideal of R is of the form RF' for an ultrafilter F of B.*

Proof. (1) Clearly, RF' forms an ideal of R. We show the second assertion. If $\mathrm{E}[r] \in F'$ then $r = \mathrm{E}[r]r \in RF'$. Conversely, let $r \overset{\text{def.}}{=} \sum_i r_i e_i \in RF'$, where $e_i \in F'$ and $r_i \in R$. Set $e \overset{\text{def.}}{=} \bigvee_i e_i$. Since F' is an ideal of B, $e \in F'$. We have $er = \sum_i r_i(ee_i) = \sum_i r_i e_i = r$. So $e \geq \mathrm{E}[r]$. Since $e \in F'$, $\mathrm{E}[r] \in F'$.

(2) Let P be a prime ideal of R. Define $F \overset{\text{def.}}{=} \{e \in B : eR \not\subseteq P\}$. We claim that F is an ultrafilter. Since $R \not\subseteq P$, $1 \in F$. For $e_1, e_2 \in B$, if $e_1 \leq e_2$ then

$Re_1 \subseteq Re_2$. So $Re_1 \not\subseteq P$ implies $Re_2 \not\subseteq P$. That is, $e_1 \in F$ implies $e_2 \in F$. Now, let $e_1, e_2 \in F$. Note that $e_1(1 - e_2), e_1e_2, e_2(1 - e_1) \in B$ are orthogonal to each other and $e_1 \vee e_2 = e_1(1 - e_2) + e_1e_2 + e_2(1 - e_1)$. By the orthogonal completeness of R,

$$Re_1 \vee e_2 = Re_1(1 - e_2) \oplus Re_1e_2 \oplus Re_2(1 - e_1).$$

The prime ideal P must include at least two of the three direct summands Re_1e_2, $Re_1(1 - e_2)$, $Re_2(1 - e_1)$. Since $e_1 \in F$, $P \not\supseteq Re_1 = Re_1(1 - e_2) \oplus Re_1e_2$. Similarly, $P \not\supseteq Re_2 = Re_2(1 - e_1) \oplus Re_1e_2$. So we must have $P \supseteq Re_1(1 - e_2) \oplus Re_2(1 - e_1)$ and $P \not\supseteq Re_1e_2$. So $e_1 \wedge e_2 = e_1e_2 \in F$. For $e \in B$, since $R = eR \oplus (1 - e)R$, either $Re \not\subseteq P$ or $R(1 - e) \not\subseteq P$. So either $e \in F$ or $1 - e \in F$. So F is an ultrafilter as claimed. By the definition of F, $P \supseteq RF'$. We have shown that any prime ideal P of R includes RF' for an ultrafilter F of B. Now, it suffices to show that RF', where F is an ultrafilter, is a prime ideal of R.

Assume that F is an ultrafilter. Suppose that $aRb \subseteq RF'$. By (8) of Lemma 2.2, $\mathrm{E}[aRb] = \mathrm{E}[a]\mathrm{E}[b]$. We well-order R as $c_0 < c_1 < \cdots$. Define $e_0 \overset{\mathrm{def.}}{=} \mathrm{E}[ac_0b]$ and inductively $e_i \overset{\mathrm{def.}}{=} \mathrm{E}[ac_ib] - \bigvee_{j<i} e_j$. Clearly, $\bigvee_i e_i = \bigvee_i \mathrm{E}[ac_ib] = \mathrm{E}[aRb]$ and these e_i are orthogonal. By the orthogonal completeness of R, there exists $c \in R$ such that $e_ic = e_ic_i$. So $e_iacb = e_iac_ib$ and hence $e_i\mathrm{E}[acb] = \mathrm{E}[e_iacb] = \mathrm{E}[e_iac_ib] = e_i\mathrm{E}[ac_ib] = e_i$. It follows that $\mathrm{E}[acb] = \bigvee_i e_i = \mathrm{E}[aRb] = \mathrm{E}[a]\mathrm{E}[b]$. Since $acb \in aRb \subseteq RF'$, we have $\mathrm{E}[a]\mathrm{E}[b] = \mathrm{E}[acb] \in F'$. Since F' is an ultrafilter, either $\mathrm{E}[a] \in F'$ or $\mathrm{E}[b] \in F'$, that is, either $a \in RF'$ or $b \in RF'$. So RF' is a prime ideal.   □

We want to expand $\mathcal{L}$ by adding predicate and function symbols. Let P be an n-ary predicate symbol and designate an n-ary predicate of R, which we also denote by P. In (2.1), $\|a = b\|$ is defined to be the greatest $e \in B$ such that $ea = eb$. We treat predicates of R in the same spirit as follows. Given an n-ary predicate P of R, we define the Boolean predicate $(a_1, \ldots, a_n) \in R^n \to \|P(a_1, \ldots, a_n)\| \in B$ by

$$\|P(a_1, \ldots, a_n)\| \overset{\mathrm{def.}}{=} \begin{cases} \text{the greatest } e \in B \text{ such that } P(ea_1, \ldots, ea_n) \text{ holds} \\ \qquad \text{if such } e \text{ exists or} \\ 0, \quad \text{otherwise.} \end{cases}$$

Set $B^* \overset{\text{def.}}{=} \{e \in B : e \neq 0\}$. So for any $e \in B^*$, if $P(ea_1, \ldots, ea_n)$ holds then $e \leq \|P(a_1, \ldots, a_n)\|$. In order to make the converse implication hold, we are thus led to the following

Definition 2.5 ([1, p. 68]). An n-ary predicate P of R is called admissible if for any $a_1, \ldots, a_n \in R$ and any $e \in B^*$,

$$(2.2) \qquad P(ea_1, \ldots, ea_n) \text{ holds iff } 0 < e \leq \|P(a_1, \ldots, a_n)\|.$$

We now consider functions. Let f be an n-ary symbol and designate an n-ary function of R, which we also denote by f. The property we impose on f is that the $(n+1)$-ary predicate $f(x_1, \ldots, x_n) = y$ is admissible. That is, for any elements $a_1, \ldots, a_n, c \in R$, there exists $u \in B$, depending on $a_1, \ldots, a_n, c$, such that for any $e \in B^*$,

$$(2.3) \qquad f(ea_1, \ldots, ea_n) = ec \text{ holds iff } 0 < e \leq u.$$

If we set c to be the element $f(a_1, \ldots, a_n)$ then (2.3) holds with $e = 1$. So for $c \overset{\text{def.}}{=} f(a_1, \ldots, a_n)$, $u \overset{\text{def.}}{=} u(a_1, \ldots, a_n, c)$ must be 1 and (2.3) says that

$$(2.4) \qquad f(ea_1, \ldots, ea_n) = ef(a_1, \ldots, a_n) \qquad \text{for any } e \in B^*.$$

Conversely, assume (2.4). We show that (2.3) holds with

$$u \overset{\text{def.}}{=} \|f(a_1, \ldots, a_n) = c\|.$$

If $0 < e \leq \|f(a_1, \ldots, a_n) = c\|$ then $ef(a_1, \ldots, a_n) = ec$. By (2.4),

$$f(ea_1, \ldots, ea_n) = ef(a_1, \ldots, a_n).$$

So $f(ea_1, \ldots, ea_n) = ec$. On the other hand, if $f(ea_1, \ldots, ea_n) = ec$ then $ef(a_1, \ldots, a_n) = ec$ by (2.4). So

$$e \leq \|f(a_1, \ldots, a_n) = c\| \overset{\text{def.}}{=} u.$$

Hence (2.3) and (2.4) are actually equivalent. We are thus led to the following

Definition 2.6 ([1, p. 59]). An n-ary function $f : R^n \to R$ of R is called admissible if it satisfies (2.4).

The definition given in [1, p. 59] also requires that

$$(2.5) \qquad f(a_1, \ldots, a_n) \in eR \qquad \text{for } a_1, \ldots, a_n \in eR \text{ and } e \in B^*.$$

But this follows immediately from (2.4).

Lemma 2.7. *If $f : R^n \to R$ is admissible then f satisfies Axiom 4. If $P \subseteq R^n$ is admissible then the Boolean predicate $(a_1, \ldots, a_n) \in R^n \mapsto \|P(a_1, \ldots, a_n)\| \in B$ satisfies Axiom 5.*

Proof. Let $a_i, b_i \in R$, where $1 \leq i \leq n$, be given. To show that the function f satisfies Axiom 4, set $e \overset{\text{def.}}{=} \bigwedge_{i=1}^n \|a_i = b_i\|$. Assume $e \neq 0$. By the admissibility of f, $f(ea_1, \ldots, ea_n) = ef(a_1, \ldots, a_n)$ and $f(eb_1, \ldots, eb_n) = ef(b_1, \ldots, b_n)$. We also have $ea_i = eb_i$ for $i = 1, \ldots, n$. So $ef(a_1, \ldots, a_n) = ef(b_1, \ldots, b_n)$. Therefore, $e \leq \|f(a_1, \ldots, a_n) = f(b_1, \ldots, b_n)\|$ if $e \neq 0$. But the inequality holds trivially if $e = 0$. So the inequality holds always and is equivalent to

$$\|a_1 = b_1 \wedge \cdots \wedge a_n = b_n \to \big(f(a_1, \ldots, a_n) = f(b_1, \ldots, b_n)\big)\| = 1.$$

So Axiom 4 holds for f. To show that the predicate P satisfies Axiom 5, set

$$e \overset{\text{def.}}{=} \bigwedge_{i=1}^n \|a_i = b_i\| \wedge \|P(a_1, \ldots, a_n)\|.$$

Assume $e \neq 0$. By the admissibility of P, $P(ea_1, \ldots, ea_n)$ holds in R. We also have $ea_i = eb_i$ for $i = 1, \ldots, n$. Therefore, $P(eb_1, \ldots, eb_n)$ also holds in R. By the admissibility of P again, $e \leq \|P(b_1, \ldots, b_n)\|$. So $e \leq \|P(b_1, \ldots, b_n)\|$ if $e \neq 0$. But if $e = 0$ then the inequality holds trivially. So we always have

$$\|a_1 = b_1 \wedge \cdots \wedge a_n = b_n \wedge P(a_1, \ldots, a_n)\| \leq \|P(b_1, \ldots, b_n)\|.$$

This is equivalent to

$$\|a_1 = b_1 \wedge \cdots \wedge a_n = b_n \to \big(P(a_1, \ldots, a_n) \to P(b_1, \ldots, b_n)\big)\| = 1.$$

Similarly, we have

$$\|a_1 = b_1 \wedge \cdots \wedge a_n = b_n \to \big(P(b_1, \ldots, b_n)\big) \to P(a_1, \ldots, a_n)\| = 1.$$

So Axiom 5 holds for the Boolean predicate

$$(a_1, \ldots, a_n) \in R^n \to \|P(a_1, \ldots, a_n)\| \in B.$$

$\square$

It is convenient to have

Definition 2.8. Let $\mathcal{L}$ be the language of ring theory adjoined by predicate symbols P's and function symbols f's. An orthogonally complete semiprime ring R is called an admissible $\mathcal{L}$-ring if for each predicate symbol P (and resp. function symbol f), there is associated an admissible predicate (and an admissible function resp.) of R of the same arity. For an admissible $\mathcal{L}$-ring R, we associate each predicate symbol P the Boolean predicate in Lemma 2.7. The resulting Boolean model is called the Boolean $\mathcal{L}$-ring R.

An admissible $\mathcal{L}$-ring R is not merely a ring. It also possesses additional structures determined by adjoined functions and predicates. So a subring of R may not inherit the $\mathcal{L}$-structure of R. But a direct summand of R does inherit a natural $\mathcal{L}$-substructure as we will see now. A direct summand of R is clearly of the form eR for some $e \in B$. Since R is orthogonally complete, eR is a subset of R for any $e \in B$ and hence forms a direct summand of R. The direct summand eR obviously has the extended centroid eC and the set eB of central idempotents. So eR is also an orthogonally complete semiprime ring over eB. By (2.5), an admissible function restricted to eR gives rise to a function of eR, which is clearly admissible. This might be the reason why (2.5) is included in [1]. Also, the restriction to eR of an admissible predicate $P \subseteq R^n$ gives rise to the admissible predicate $P \cap (eR)^n$ of eR, since $\|P(ea_1, \ldots, ea_n)\| \in eB$ for $a_1, \ldots, a_n \in R$. We thus call eR an $\mathcal{L}$-subring of R.

The admissibility of R can also be understood in terms of direct summands of R as follows. To be suggestive, we call ea the projection of $a \in R$ in the direct summand eR. The Boolean value $\|a = b\|$ is the greatest $e \in B$ such that the projections of a, b in the direct summand eR are equal. More generally, for an admissible predicate $P \subseteq R^n$, $\|P(a_1, \ldots, a_n)\|$ is the greatest $e \in B$ such that the ordered tuple of projections of a_i in the direct summand eR satisfies the restriction $P \cap (eR)^n$ of P. For an admissible function $f : R^n \to R$, $\|f(a_1, \ldots, a_n) = c\|$ is the greatest $e \in B$ such that the restriction of f to the direct summand eR sends the ordered tuple of projections of a_i to the projection of c. So restrictions of predicates and functions to direct summands completely determine their Boolean values. These observations establish the following important connection.

Lemma 2.9. *Let R be an admissible $\mathcal{L}$-ring. If $F \stackrel{\text{def.}}{=} \{f \in B : f \geq e\}$, the principal filter defined by $e \in B$, then the reduced product R_F and the $\mathcal{L}$-subring eR are isomorphic $\mathcal{L}$-structures by the map*

$$a_F \in R_F \leftrightarrow ea \in eR.$$

Proof. We check that the asserted isomorphism is well-defined: For $a, b \in R$,

$$a_F = b_F \stackrel{\text{def}}{\Longleftrightarrow} \|a = b\| \in F \Longleftrightarrow e \leq \|a = b\| \Longleftrightarrow ea = eb,$$

where the first equivalence follows from the the definition of the filtered structure R_F, where the second follows from the definition of the filter F, and where the third follows from the definition of $\|a = b\|$. Let the predicate symbol $P \in \mathcal{L}$ designate an admissible n-ary predicate, which we also denote by P. We have

$$P(a_{1,F}, \ldots, a_{n,F}) \text{ holds in } R_F \Leftrightarrow \|P(a_1, \ldots, a_n)\| \in F$$
$$\Leftrightarrow e \le \|P(a_1, \ldots, a_n)\|$$
$$\Leftrightarrow P(ea_1, \ldots, ea_n) \text{ holds in } R$$
$$\Leftrightarrow P(ea_1, \ldots, ea_n) \text{ holds in the substructure } eR.$$

Let the function symbol $f \in \mathcal{L}$ designate an admissible n-ary function, which we also denote by f. We have similarly

$$f(a_{1,F}, \ldots, a_{n,F}) = b_F \text{ in } R_F \Leftrightarrow \|f(a_1, \ldots, a_n) = b\| \in F$$
$$\Leftrightarrow e \le \|f(a_1, \ldots, a_n) = b\|$$
$$\Leftrightarrow ef(a_1, \ldots, a_n) = eb \text{ holds in } R$$
$$\Leftrightarrow f(ea_1, \ldots, ea_n) = eb \text{ holds in } eR.$$

$\square$

We are now seeking quotient rings of an admissible $\mathcal{L}$-ring R which inherits the $\mathcal{L}$-structure.

Definition 2.10. Let I be an ideal of R. Denote the natural images of $a \in R$ in R/I by $\bar{a}$. Write $a =_I b$ for $a, b \in R$ to denote $a - b \in I$ or equivalently to denote $\bar{a} = \bar{b}$.

(1) Given an n-ary predicate $P \subseteq R^n$, we define the quotient predicate $\overline{P}$ of R/I as follows: For $a_1, \ldots, a_n \in R$, $(\bar{a}_1, \ldots, \bar{a}_n) \in \overline{P}$ iff there exist $b_i \in R$ with $a_i \equiv_I b_i$ for $1 \le i \le n$ such that $(b_1, \ldots, b_n) \in P$.

(2) Given an n-ary function $f : R^n \to R$, we define the quotient function $\overline{f} : (R/I)^n \to R/I$ by setting $\overline{f}(\bar{a}_1, \ldots, \bar{a}_n) \overset{\text{def.}}{=} \overline{f(a_1, \ldots, a_n)} \in R/I$ for $a_1, \ldots, a_n \in R$. The quotient function $\overline{f}$ is well-defined iff the function f is I-invariant in the sense that for any $a_i, b_i \in R$, $a_i \equiv_I b_i$ for $1 \le i \le n$ implies $f(a_1, \ldots, a_n) \equiv_I f(b_1, \ldots, b_n)$.

The necessary and sufficient condition for the quotient function $\overline{f}$ to be well-defined is obvious. We show that admissible functions are I-invariant for filter ideals I. *Reason*: In the notation of Lemma 2.4, let $I \overset{\text{def.}}{=} RF'$, where F is a filter of B. If $a_i \equiv_I b_i$ for $1 \le i \le n$ then there exists $0 \ne e \in F$ such that $ea_i = eb_i$ for $1 \le i \le n$ by (1) of Lemma 2.4. By the admissibility of f, $ef(a_1, \ldots, a_n) = f(ea_1, \ldots, ea_n) = f(eb_1, \ldots, eb_n) = ef(b_1, \ldots, b_n)$. So $f(a_1, \ldots, a_n) \equiv_I f(b_1, \ldots, b_n)$, as asserted.

In universal algebras, predicates are treated very much like functions and are required to be I-invariant in the following sense. In the notation of Definition 2.10, a predicate $P \subseteq R^n$ is called I-invariant if for any $a_i, b_i \in R$ with $a_i \equiv_I b_i$ for $i = 1, \ldots, n$, we have

$$(a_1, \ldots, a_n) \in P \text{ iff } (b_1, \ldots, b_n) \in P.$$

The quotient predicate $\overline{P}$ of R/I for an I-invariant predicate $P \subseteq R^n$ is then defined by the following

$$(\overline{a}_1, \ldots, \overline{a}_n) \in \overline{P} \text{ iff } (a_1, \ldots, a_n) \in P \text{ for } a_1, \ldots, a_n \in R.$$

The truth of $\overline{P}(\overline{a}_1, \ldots, \overline{a}_n)$ is independent of the choice of a_i by the I-invariance of P. This definition is clearly equivalent to Definition 2.10 if the predicate P is I-invariant. The quotient predicate $\overline{P}$ of an I-invariant predicate P does preserve many nice properties of P. If we want to work in this context, we must be looking for predicates which are I-invariant for *all* filter ideals I. Unfortunately, such predicates of R are either R^n or $\varnothing$ if $B \supsetneq \{0, 1\}$. *Reason*: Suppose that $\varnothing \neq P \subseteq R^n$ is such an n-ary predicate. Pick $(a_1, \ldots, a_n) \in P$ and $e \in B$ such that $0 < e < 1$. Given any $b_1 \in R$, define $c \stackrel{\text{def.}}{=} ea_1 + (1 - e)b_1$. Since $a_{1,F} = c_F$ for any filter F with $e \in F$, we have $(c, a_2, \ldots, a_n) \in P$ by the assumed RF'-invariance of P. Similarly, since $c_F = b_{1,F}$ for any filter F with $1 - e \in F$, we have $(b_1, a_2, \ldots, a_n) \in P$ by the assumed RF'-invariance of P again. Proceeding in this manner, we see that $(b_1, b_2, a_3, \ldots), \ldots, (b_1, b_2, \ldots, b_n) \in P$ for any $b_1, b_2, \ldots \in R$. Since $b_1, b_2, \ldots$ are arbitrary, we have $P = R^n$, as asserted. Therefore, the invariance for filtered ideals for predicates is too restrictive to be useful for our purpose and must be given up.

Definition 2.11. Let R be an admissible $\mathcal{L}$-ring. For a filter ideal I, the quotient $\mathcal{L}$-structure given in Definition 2.10 is denoted by R/I.

Lemma 2.12. *Let R be an admissible $\mathcal{L}$-ring and F, a filter of B. Set*

$$F' \stackrel{\text{def.}}{=} \{1 - e : e \in F\} \quad \text{and} \quad I \stackrel{\text{def.}}{=} RF'.$$

Let $\overline{a}$ denote the natural image in R/I of $a \in R$. The reduced product R_F is then isomorphic to the quotient $\mathcal{L}$-structure R/RF' via the map $a_F \in R_F \leftrightarrow \overline{a} \in R/I$ for $a \in R$.

Proof. Clearly, $a_F \in R_F \leftrightarrow \overline{a} \in R/RF'$ gives a bijection between R_F and R/RF'. Let the predicate symbol $P \in \mathcal{L}$ designate an n-ary admissible predicate of R which we also denote by P. Let $a_1, \ldots, a_n \in R$ be given. We have the following equivalence

$\overline{P}(\overline{a}_1, \ldots, \overline{a}_n)$ holds in the quotient structure R/I

$\Leftrightarrow$ for some $b_i \in R$ with $a_i \equiv_I b_i$ for all i, $P(b_1, \ldots, b_n)$ holds in R

$\Leftrightarrow$ for some $e \in F$ and $b_i \in R$ with $ea_i = eb_i$ for all i, $P(b_1, \ldots, b_n)$ holds in R

$\Leftrightarrow$ for some $e \in F$, $P(ea_1, \ldots, ea_n)$ holds in R

$\Leftrightarrow$ for some $e \in F$, $e \leq \|P(a_1, \ldots, a_n)\|$

$\Leftrightarrow \|P(a_1, \ldots, a_n)\| \in F$

$\Leftrightarrow P(a_{1,F} \ldots, a_{n,F})$ holds in R_F.

For $\Rightarrow$ of third equivalence, since $e \in F, e \neq 0$. By the admissibility of P, the truth of $P(b_1, \ldots, b_n)$ in R implies $\|P(b_1, \ldots, b_n)\| = 1$. So $0 < e \leq \|P(b_1, \ldots, b_n)\|$. By the admissibility of P again, $P(eb_1, \ldots, eb_n)$ also holds in R. But $ea_i = eb_i$. So $P(ea_1, \ldots, ea_n)$ holds in R. For $\Leftarrow$ of third equivalence, we merely take $b_i \overset{\text{def.}}{=} ea_i$.

Let the function symbol $f \in \mathcal{L}$ designate an n-ary admissible function of R which we also denote by f. We have similarly

$\overline{f}(\overline{a}_1, \ldots, \overline{a}_n) = \overline{b}$ in the quotient structure R/I

$\Leftrightarrow \overline{f(a_1, \ldots, a_n)} = \overline{b}$ in R/I by the definition of $\overline{f}$

$\Leftrightarrow f(a_1, \ldots, a_n) - b \in I$

$\Leftrightarrow \|f(a_1, \ldots, a_n) = b\| \in F$ since $I \overset{\text{def.}}{=} RF'$

$\Leftrightarrow f(a_{1,F}, \ldots, a_{n,F}) = b_F$ in R_F by the definition of R_F.

Note that we only use the I-invariance of f in the above   □

We have come to the main result of this paper.

Theorem (Beidar and Mikhalëv [1]). *Let R be an admissible $\mathcal{L}$-ring. Let θ be a sentence of $\mathcal{L}$.*

(1) *Assume that θ is Horn. If θ holds on R/P for any minimal prime ideal P of R, then θ holds on any direct summand of R.*

(2) *Assume that $\neg\theta$ is Horn. If θ holds on any direct summands of R, then θ holds on R/P for any minimal prime ideal P of R.*

Proof. (1) Assume that θ holds on R/P for any minimal prime ideal P of R. If $\|\theta\| \neq 1$ then we extend $1 - \|\theta\|$ to an ultrafilter F of B. By Łôs Theorem, $\neg\theta$

holds in the ultraproduct R_F and hence in the quotient structure R/RF', since R_F is isomorphic to R/RF' by Lemma 2.12. But RF' is a minimal prime ideal of R by (2) of Lemma 2.4. This contradicts our assumption that θ holds on R/P for any minimal prime ideal P of R. So $\|\theta\| = 1$. Now, θ is Horn. By Horn Theorem, θ holds in any reduced products, particularly in R_F, where F is a principal filter, say $F \overset{\text{def.}}{=} \{f \in B : f \geq e\}$. By Lemma 2.9, R_F is isomorphic to the direct summand eR as $\mathcal{L}$-structures. So θ holds in eR. Since $e \in B$ is arbitrary. (1) is proved.

(2) Assume that $\neg\theta$ is Horn. If $\|\neg\theta\| > 0$, then the Horn sentence $\neg\theta$ holds in the reduced product R_F, where F is the principal filter $F \overset{\text{def.}}{=} \{e \in B : e \geq \|\neg\theta\|\}$. But R_F is isomorphic to the direct summand $\|\neg\theta\|R$ by Lemma 2.9. This contradicts our assumption that θ holds in any direct summands of R. So $\|\neg\theta\| = 0$ and hence $\|\theta\| = 1$. By Łoś Theorem, θ holds on any ultraproduct R_F, which is isomorphic to the quotient $\mathcal{L}$-structure R/RF' by Lemma 2.12. By Lemma 2.4, any minimal prime ideal is of the form RF' for an ultrafilter F of B. (2) is proved. $\qquad\square$

3. A generalization

We have thus proved the main result of Beidar and Mikhalëv's theory for semi-prime rings. Now we will look this theory from the view point of general Boolean valued models. We would like to get a generalization together with new insights. Let us go back for a moment to general Boolean valued models over a complete Boolean algebra B.

Clearly, an orthogonal family $e_i \in B$ is maximal if and only if $\bigvee_i e_i = 1$. A maximal orthogonal family $e_i \in B$ is also called dense because it is dense in B with respect to the order topology of B defined by taking $\{e \in B : e \leq p\}$ as basic open sets. (See [4, p. 153].) Another motivation is that the orthogonal family $\{e_i\}$ is maximal if and only if the ideal it generates is dense in Utumi's sense. (See [5].) Let $\mathcal{A}$ be a Boolean valued model over B. We say that $a, b \in A$ are almost equal and write $a \overset{\text{a.e.}}{=} b$ if $\|a = b\| = 1$. Let $e_i \in B$ form a maximal orthogonal family. By mixing principle, for any $a_i \in A$, there exists $a \in A$ such that $\|a = a_i\| \geq e_i$. We claim that any two such a's are almost equal. *Reason*: Suppose that $b \in A$ also satisfies $\|b = a_i\| \geq e_i$ for all i. Then $\|a = b\| \geq \|a = a_i \wedge b = a_i\| \overset{\text{def.}}{=} \|a = a_i\|\|b = a_i\| \geq e_i$. This is true for all i. So $\|a = b\| \geq \bigvee_i e_i = 1$. So $a \overset{\text{a.e.}}{=} b$, as asserted. It is convenient to have

Definition 3.1. If $e_i \in B$ forms a maximal orthogonal family, then we denote the element a such that $\|a = a_i\| \geq e_i$ by $a \overset{\text{a.e.}}{=} \sum_i^\perp e_i \cdot a_i$.

Note that the mixing principle deals with the equality predicate only and has nothing to do with other functions or Boolean predicates. But via Axioms 4 and 5, mixing principle imposes the following orthogonal linearity on all Boolean predicates.

Lemma 3.2. *If $\mathcal{A}$ is a Boolean valued model satisfying the mixing principle, then for any maximal orthogonal family $e_i \in B$, for any $a_i \in A$ and for any formula $\varphi(x)$,*

$$\left\| \varphi\left(\sum_i^\perp e_i \cdot a_i\right) \right\| = \left\| \sum_i^\perp e_i \cdot \varphi(a_i) \right\|.$$

Proof. Write $a \overset{\text{a.e.}}{=} \sum_i^\perp e_i \cdot a_i$ for short. By Lemma 1.3, we have

$$\|(a = a_i) \to (\varphi(a_i) \equiv \varphi(a))\| = 1.$$

So $\|a = a_i\| \|\varphi(a)\| = \|a = a_i\| \|\varphi(a_i)\|$. Since $\|a = a_i\| \geq e_i$, we have $e_i \|\varphi(a)\| = e_i \|\varphi(a_i)\|$. Summing up all these and using $\sum_i^\perp e_i = 1$, we have

$$\|\varphi(a)\| = \sum_i^\perp e_i \|\varphi(a)\| = \sum_i^\perp e_i \|\varphi(a_i)\|,$$

as asserted. $\qquad\qquad\qquad\qquad\qquad\qquad\qquad\qquad\qquad\qquad\qquad\qquad\square$

In a Boolean valued model, the equality axioms 4 and 5 actually impose severe restrictions on functions and Boolean predicates respectively. Let us analyze these restrictions with the following

Definition 3.3. (1) A unary Boolean predicate $\tilde{P} : A \to B$ is said to preserve almost equality if $a \overset{\text{a.e.}}{=} b$ implies $\tilde{P}(a) = \tilde{P}(b)$. We call a unary Boolean predicate $\tilde{P} : A \to B$ orthogonally linear if it preserves the almost equality and for any $e \in B$ and $a, b \in A$,

$$\tilde{P}(e \cdot a + (1 - e) \cdot b) = e\tilde{P}(a) + (1 - e)\tilde{P}(b).$$

An n-ary Boolean predicate $\tilde{P} : A^n \to B$ is said to preserve the almost equality (or to be orthogonally linear resp.) if it preserves the almost equality (or is orthogonally linear resp.) in each argument, that is, if for any $1 \leq i \leq n$, the unary Boolean predicate obtained from $\tilde{P}$ by assigning arbitrarily fixed elements in A to all but the i-th argument preserves the almost equality (or is orthogonally linear resp.).

(2) An n-ary function $\tilde{f} : A^n \to A$ is said to preserve the almost equality if for any $a_i, b_i \in A$, $a_i \overset{\text{a.e.}}{=} b_i$ for $i = 1, \ldots, n$ implies

$$\tilde{f}(a_1, \ldots, a_n) \overset{\text{a.e.}}{=} \tilde{f}(b_1, \ldots, b_n).$$

An n-ary function $\tilde{f} : A^n \to A$ is said to be orthogonally linear if it preserves the almost equality and if the $(n+1)$-ary Boolean predicate

$$(a_1, \ldots, a_n, b) \in A^{n+1} \to \|f(a_1, \ldots, a_n) = b\|$$

is orthogonally linear in the first n arguments.

Note that we have assumed the preservation of almost equality in the definition of orthogonal linearity, because the element $c \overset{\text{a.e.}}{=} e \cdot a + (1 - e) \cdot b$ there is only unique up to within almost equality. With the notion of orthogonal linearity, we analyze the restrictions imposed by Axioms 4 and 5 as follows.

Lemma 3.4. *Let A be a Boolean valued model satisfying the mixing principle. Let us adjoin to $\mathcal{L}$ a new n-ary predicate symbol P interpreted as a Boolean predicate $\tilde{P} : A^n \to B$ (resp. a new n-ary function symbol f interpreted as a function $\tilde{f} : A^n \to A$). Then Axiom 5 (resp. Axiom 4) for P (resp. for f) holds if and only if $\tilde{P}$ (resp. $\tilde{f}$) is orthogonally linear.*

Proof. For the only-if part, assume Axiom 5 (resp. Axiom 4) holds for $\tilde{P}$ (resp. $\tilde{f}$). Then A adjoined by $\tilde{P}$ (resp. $\tilde{f}$) forms a Boolean valued model for the language $\mathcal{L}$ adjoined by P (resp. by f). Axiom 5 (resp. Axiom 4) implies the preservation of almost equality for $\tilde{P}$ (resp. for $\tilde{f}$). The orthogonal linearity follows from Lemma 3.2.

For the if part, we first consider the Boolean predicate $\tilde{P}$ with $n = 1$. Given $a, b \in A$, set $e \overset{\text{def.}}{=} \|a = b\|$. Define $c \overset{\text{a.e.}}{=} e \cdot b + (1 - e) \cdot a$. By the orthogonal linearity of $\tilde{P}$,

$$\tilde{P}(c) = \tilde{P}(e \cdot b + (1 - e) \cdot a) = e\tilde{P}(b) + (1 - e)\tilde{P}(a).$$

Multiplying this by e, we have $e\tilde{P}(c) = e\tilde{P}(b)$. On the other hand, $\|c = b\| \geq e$ by the definition of c. By Axiom 3,

$$\|c = a\| \geq \|(c = b) \wedge (b = a)\| \overset{\text{def.}}{=} \|c = b\| \|b = a\| \geq e^2 = e.$$

Also, $\|c = a\| \geq 1 - e$. So $\|c = a\| \geq e \vee (1 - e) = 1$, that is, $c \overset{\text{a.e.}}{=} a$. But the orthogonal linearity of $\tilde{P}$ assumes the preservation of almost equality in the definition. So $\tilde{P}(c) = \tilde{P}(a)$. Together with $e\tilde{P}(c) = e\tilde{P}(b)$ obtained before, we have $e\tilde{P}(a) = e\tilde{P}(b)$, that is, $e \leq \|\tilde{P}(a) \equiv \tilde{P}(b)\|$. Since $e \overset{\text{def.}}{=} \|a = b\|$, we have

$$\|(a = b) \rightarrow (\tilde{P}(a) \equiv \tilde{P}(b))\| = 1.$$

This holds for all $a, b \in A$. So Axiom 5 for $\tilde{P}$ follows.

Now, assume that $\tilde{P}$ is n-ary. Let a_i, b_i, $i = 1, \ldots, n$, be given. Apply the result for the unary Boolean predicate to each argument of $\tilde{P}$ successively. We have

$$\|(a_1 = b_1)\| \leq \|\tilde{P}(a_1, a_2, a_3, \ldots) \equiv \tilde{P}(b_1, a_2, a_3, \ldots)\|,$$
$$\|(a_2 = b_2)\| \leq \|\tilde{P}(b_1, a_2, a_3, \ldots) \equiv \tilde{P}(b_1, b_2, a_3, \ldots)\|,$$
$$\|(a_3 = b_3)\| \leq \|\tilde{P}(b_1, b_2, a_3, \ldots) \equiv \tilde{P}(b_1, b_2, b_3, \ldots)\|,$$
$$\vdots$$

By Axioms 1-3, the formula

$$\|(a_1 = b_1 \wedge \cdots \wedge a_n = b_n) \rightarrow (\tilde{P}(a_1, a_2, \ldots) \equiv \tilde{P}(b_1, b_2, \ldots))\| = 1.$$

This is true for all $a_i, b_i \in A$. So Axiom 5 holds for $\tilde{P}$, as asserted.

We now consider the n-ary function $\tilde{f} : A^n \rightarrow A$. By the assumption, the $(n+1)$-Boolean predicate

$$(a_1, \ldots, a_n, b) \in A^{n+1} \rightarrow \|\tilde{f}(a_1, \ldots, a_n) = b\|$$

is orthogonally linear in the first n arguments. Applying the result for Boolean predicates to this, we have

$$\|(a_1 = b_1) \wedge \cdots \wedge (a_n = b_n)\| \leq \|(\tilde{f}(a_1, \ldots, a_n) = b) \equiv (\tilde{f}(b_1, \ldots, b_n) = b)\|.$$

But

$$\|(\tilde{f}(a_1, \ldots, a_n) = b) \equiv (\tilde{f}(b_1, \ldots, b_n) = b))\|$$
$$\leq \|\tilde{f}(a_1, \ldots, a_n) = \tilde{f}(b_1, \ldots, b_n)\|.$$

So

$$\|(a_1 = b_1) \wedge \cdots \wedge (a_n = b_n)\| \leq \|\tilde{f}(a_1, \ldots, a_n) = \tilde{f}(b_1, \ldots, b_n)\|.$$

Therefore, f satisfies Axiom, as asserted. $\qquad\qquad\qquad\qquad\qquad\square$

We now come back to semiprime rings. With the Boolean value assigned to the equality $=$ in (2.1), the ring R is a Boolean valued model for the simplest language with $=$ only. We want to expand $\mathcal{L}$ by adding function symbols and predicate symbols. To obtain Boolean valued models, functions adjoined must satisfy Axiom 4, that is, they must be orthogonally linear by Lemma 3.4. Admissible functions satisfy Axiom 4 by Lemma 2.7 and hence are orthogonally linear by Lemma 3.4. (A direct proof of this is given in Lemma 3.5 below.) Any constant function is clearly orthogonally linear but is not admissible unless it is 0. So orthogonal linearity is stronger than admissibility. Actually, they do not differ very much as the following shows

Lemma 3.5. *A function* $f : R^n \to R$ *is admissible iff it is orthogonally linear and satisfies* $f(0, \ldots, 0) = 0$.

Proof. We have seen that admissible functions are orthogonally linear by Lemmas 2.7 and 3.4. We give a direct argument as follows

Claim 1. If $f : R^n \to R$ is admissible, then $f(0, \ldots, 0) = 0$.

Reason. By the admissibility of f, for any $e \in B^*$,

$$ef(0, \ldots, 0) = f(e0, \ldots, e0) = f(0, \ldots, 0),$$

that is, $(1 - e)f(0, \ldots, 0) = 0$. Since $1 - e \in B^*$ also, replacing e by $1 - e$ in the above, we have similarly $\big(1 - (1 - e)\big)f(0, \ldots, 0) = 0$, that is, $ef(0, \ldots, 0) = 0$. So $f(0, \ldots, 0) = ef(0, \ldots, 0) + (1 - e)f(0, \ldots, 0) = 0$, as is claimed.

Claim 2. If $f : R^n \to R$ is admissible, then for any $e \in B$ and $a_1, \ldots, a_n \in R$,

$$ef(a_1, \ldots, a_n) = ef(ea_1, a_2, \ldots, a_n).$$

Reason. If $e = 0$ then both sides are 0. If $e = 1$ the equality holds trivially. So assume $e \neq 0$ and $e - 1 \neq 0$. By the admissibility, we have

$$
\begin{aligned}
ef(a_1, \ldots, a_n) &= f(ea_1, \ldots, ea_n) \\
&= f(e^2 a_1, ea_2, \ldots, e_n) = ef(ea_1, a_2, \ldots, a_n),
\end{aligned}
$$

as is claimed.

We are ready for a direct proof of the only-if part. Assume that $f : R^n \to R$ is admissible. We have $f(0, \ldots, 0) = 0$ by Claim 1. We show the orthogonal linearity of f. Given $a_1 = eb + (1 - e)c$, we have $b = ea_1$ and $c = (1 - e)a_1$. By Claim 2, we have

$$ef(a_1, a_2, \ldots) = ef(ea_1, a_2, \ldots) = ef(b, a_2, \ldots) \quad \text{and}$$
$$(1 - e)f(a_1, a_2, \ldots) = (1 - e)f((1 - e)a_1, a_2, \ldots) = (1 - e)f(c, a_2, \ldots).$$

The sum of the two equalities gives the orthogonal linearity of f in the first argument

$$f(eb + (1 - e)c, a_2, \ldots) = f(a_1, a_2, \ldots) = ef(b, a_2, \ldots) + (1 - e)f(c, a_2, \ldots).$$

The orthogonal linearity in other arguments follows analogously. So f is orthogonally linear. This proves the only-if part.

The if part is more complicate. An n-ary function g, $n \geq 1$, is called good if g assumes the value 0 when any one of its arguments is set to be 0.

Claim 3. The following are equivalent for a good function $g : R^n \to R$ with $n \geq 1$:

(i) g is admissible.

(ii) g is orthogonally linear.

(iii) For any $e \in B$ and $a_1, \ldots, a_n \in R$,

$$eg(a_1, \ldots, a_n) = g(ea_1, a_2, a_3, \ldots) = g(a_1, ea_2, a_3, \ldots)$$
$$= \cdots = g(a_1, \ldots, a_{m-1}, ea_m).$$

Reason. By the only-if part of this lemma, (i) implies (ii). Clearly, (iii) implies (i). It suffices to show that (ii) implies (iii). So assume (ii). By the orthogonal linearity of g in the first argument, we have for $e \in B$ and $a_1, \ldots, a_n \in R$,

$$g(ea_1, a_2, \ldots) = g(ea_1 + (1 - e)0, a_2, \ldots)$$
$$= eg(a_1, a_2, \ldots) + (1 - e)g(0, a_2, \ldots) = eg(a_1, a_2, \ldots),$$

where the last equality follows from $g(0, a_2, \ldots) = 0$ by goodness of g. The rest of arguments can be treated similarly. So (iii) holds, as claimed.

With Claim 3, good functions establish a bridge between orthogonally linear functions and admissible functions. But the goodness of a function is sensitive to the variables it involves. For example, even though $g(x)$ is good, the function $h(x, y) \stackrel{\text{def.}}{=} g(x)$ may not be good, because $h(x, 0) = g(x)$ may not be 0. In the following, we decompose any function as a sum of good functions plus a constant. We have to be careful about what variables a function involves.

Claim 4. Let $f(x_1, \ldots, x_n)$ be an n-ary function of R. Then for each subset S of $\{x_1, \ldots, x_n\}$, there exists a unique function g_S in the indeterminates $x_i \in S$ such that (i) f is the sum of all such g_S and such that (ii) g_S is good for $S \neq \varnothing$.

Reason. Let $h(y_1, \ldots, y_m)$ be an m-ary function in $y_1, \ldots, y_m$. Given a subset of $S \subseteq \{y_1, \ldots, y_m\}$, let h_S be the function obtained from h by setting $y_i \notin S$ to be 0. Note that h_S is a function in the variables $y_i \in S$. So $h_{\{y_1, \ldots, y_m\}}$ is the

m-ary function $h(y_1, \ldots, y_m)$ and $h_\varnothing$ is a 0-ary function, namely, the constant $h(0, \ldots, 0)$. Define

$$\overline{h}(y_1, \ldots, y_m) \stackrel{\text{def.}}{=} \sum_S (-1)^{m-|S|} h_S.$$

By setting $y_1 = 0$, h_S becomes $h_{S-\{y_1\}}$ if $y_1 \in S$ and h_S remains the same if $y_1 \notin S$. So

$$\overline{h}(0, y_2, \ldots, y_m) = \sum_{y_1 \in S} (-1)^{m-|S|} h_{S-\{y_1\}} + \sum_{y_1 \notin S} (-1)^{m-|S|} h_S = 0,$$

where the last equality follows since each S with $y_1 \in S$ corresponds to a unique S' with $y_1 \notin S'$ simply by setting $S' \stackrel{\text{def.}}{=} S - \{y_1\}$. The same argument works for setting other $y_i = 0$ for $i \neq 1$. So $\overline{h}(y_1, \ldots, y_m)$ is good. Now, we go back to our function $f(x_1, \ldots, x_n)$. For any subset S of $\{x_1, \ldots, x_n\}$, we define our desired $g_S \stackrel{\text{def.}}{=} \overline{f_S}$. So g_S is a good function in the variables in S. We check that $f = \sum_S g_S$, where the summation ranges over all subsets of $\{x_1, \ldots, x_n\}$, as follows. Given $T \subseteq \{x_1, \ldots, x_n\}$, f_T occurs in those $g_S \stackrel{\text{def.}}{=} \overline{f_S}$ for $S \supseteq T$ and has the coefficient $(-1)^{|S|-|T|}$ in $g_S \stackrel{\text{def.}}{=} \overline{f_S}$. Given $0 \leq k \leq n - |T|$, sets S such that $|S| = |T| + k$ and such that $S \supseteq T$ are obtained by adjoining to T with k elements from $\{x_1, \ldots, x_n\} - T$. So there are $\binom{n-|T|}{k}$ such sets S. The coefficient of f_T in $\sum_S g_S$ is thus

$$\sum_{k=0}^{n-|T|} \binom{n-|T|}{k} (-1)^{|S|-|T|} = \sum_{k=0}^{n-|T|} \binom{n-|T|}{k} (-1)^k = \begin{cases} 0 & \text{if } n > |T|, \\ 1 & \text{if } n = |T|. \end{cases}$$

So $f = \sum_S g_S$ follows. For the uniqueness of g_S, we set all x_i to be 0 in the equality $f = \sum_S g_S$ and, by the goodness of g_S for $S \neq \varnothing$, we obtain $f(0, \ldots, 0) = g_\varnothing$. Then we set all $x_2 = \ldots = x_n = 0$ and obtain

$$f(x_1, 0, \ldots, 0) = g_{\{x_1\}} + f(0, \ldots, 0).$$

So $g_{\{x_1\}} = f(x_1, 0, \ldots, 0) - f(0, \ldots, 0)$. We can obtain the similar expressions for other $g_{\{x_i\}}$. Proceeding in this manner, the uniqueness of g_S follows as claimed.

We are now ready for the if part. Assume that f is orthogonally linear and satisfies $f(0, \ldots, 0) = 0$. By Claim 4, $f = \sum_{S \neq \varnothing} g_S$. The orthogonal linearity of f implies that of f_S and hence that of g_S constructed in Claim 4. By Claim 3, all g_S with $S \neq \varnothing$ are admissible and hence so is their sum $f = \sum_{S \neq \varnothing} g_S$. $\qquad\square$

We can also state Lemma 3.5 by saying that a function $f(x_1, \ldots, x_n)$ is orthogonally linear iff the function $g \overset{\text{def.}}{=} f(x_1, \ldots, x_n) - f(0, \ldots, 0)$ is admissible.

So we can add orthogonally linear functions to R and the resulting structure with the Boolean predicate $=$ defined by (2.1) forms a Boolean valued model. But it is crucially important here that reduced products should be isomorphic to some algebraic constructions such as substructures or quotient structures. Orthogonally linear functions can be constants which may not be in any direct summands. For example, if R possess 1 then 1 is not in any proper direct summand eR in spite that eR has its own multiplication identity e. If we add a constant symbol c_0 in $\mathcal{L}$ to denote 1 then eR is not a $\mathcal{L}$-subring of R. To allow orthogonally linear functions, we have to give up Lemma 2.9. Instead, we require our reduced products to be isomorphic to the corresponding quotient structures as in Lemma 2.12. As noted in the proof of Lemma 2.12, only the invariance under filter ideals is needed for functions there. This is true for orthogonally linear functions by the following.

Lemma 3.6. *Orthogonally linear functions of R are invariant under filter ideals.*

Proof. Let $f : R^n \to R$ be orthogonally linear. Let $e \in B$ and $a_i, b_i \in R$ be such that $ea_i = eb_i$ for $i = 1, \ldots, n$. It suffices to show that $\|f(b_1, b_2 \ldots) = f(a_1, a_2, \ldots)\| \geq e$. Set $f(a_1, \ldots, a_n) = c$. So $\|f(a_1, \ldots, a_n) = c\| = 1$. Write $b_1 = eb_1 + (1 - e)b_1 = ea_1 + (1 - e)b_1$. By the orthogonal linearity of f,

$$\|f(b_1, a_2, \ldots) = c\| = \|f(ea_1 + (1 - e)b_1, a_2, \ldots) = c\|$$
$$= e\|f(a_1, a_2, \ldots) = c\| + (1 - e)\|f(b_1, a_2, \ldots)\|$$
$$= e + (1 - e)\|f(b_1, a_2, \ldots)\| \geq e.$$

That is, $\|f(b_1, a_2, \ldots) = c\| \geq e$. By substituting $c = f(a_1, \ldots, a_n)$, we have $\|f(a_1, a_2, \ldots) = f(b_1, a_2, \ldots)\| \geq e$. Similarly,

$$\|f(b_1, a_2, a_3, \ldots) = f(b_1, b_2, a_3, \ldots)\| \geq e,$$

and so on. All these together imply $\|f(b_1, \ldots, b_n) = f(a_1, \ldots, a_n)\| \geq e$. $\qquad\square$

Now, we come to the problem of adding new predicate symbols to our language. There is a general construction of Boolean predicates from predicates. We describe this by vector notation for simplicity.

Definition 3.7. Let $\mathcal{A}$ be a Boolean valued model satisfying mixing principle.

(1) For any $\vec{a}_i \overset{\text{def.}}{=} (a_1^{(i)}, \ldots, a_n^{(i)}) \in A^n$ and for any maximal orthogonal subset $\{e_i\}$ of B, we define $\sum_i^\perp e_i \vec{a}_i$ to be any $\vec{a} = (a_1, \ldots, a_n) \in A^n$ such that $a_j \overset{\text{a.e.}}{=} \sum_i^\perp e_i a_j^{(i)}$ for $j = 1, \ldots, n$.

(2) An n-ary predicate $P \subseteq A^n$ is called orthogonally complete if for any $\vec{a} \in A^n$, for any $\vec{a}_i \in P$ and for any maximal orthogonal subset $\{e_i\}$ of B, $\vec{a} \overset{\text{a.e.}}{=} \sum_i^{\perp} e_i \vec{a}_i$ implies $\vec{a} \in P$. The orthogonal completion of a given $P \subseteq A^n$ is defined to be the smallest orthogonally complete set including P and is obviously equal to the set of all $\sum_i^{\perp} e_i \vec{a}_i$, where $\vec{a}_i \in P$ and where $\{e_i\}$ is a maximal orthogonal subset $\{e_i\}$ of B.

(3) For $\vec{a} \overset{\text{def.}}{=} (a_1, \ldots, a_n)$, $\vec{b} \overset{\text{def.}}{=} (b_1, \ldots, b_n) \in A^n$, we define

$$\|\vec{a} = \vec{b}\| \overset{\text{def.}}{=} \|a_1 = b_1\| \wedge \cdots \wedge \|a_n = b_n\|.$$

(4) Let $P \subseteq A^n$ be a given n-ary predicate. We define the n-ary Boolean predicate $\tilde{P} : R^n \to B$ and the n-ary predicate $\widehat{P}$ by

$$\tilde{P}(\vec{a}) \overset{\text{def.}}{=} \bigvee_{\vec{b} \in P} \|\vec{a} = \vec{b}\| \quad \text{and} \quad \widehat{P} \overset{\text{def.}}{=} \{\vec{a} \in A^n : \|\tilde{P}(\vec{a})\| = 1\}.$$

We call $\tilde{P}$ the Boolean predicate induced by P.

Lemma 3.8. *In the notation of Definition 3.7, we have the following*

(1) *The Boolean predicate $\tilde{P}$ satisfies Axiom 5.*
(2) *$\widehat{P}$ also induces the Boolean predicate $\tilde{P}$ and is the maximal such subset of A^n.*
(3) *$\widehat{P}$ is the orthogonal completion of P.*
(4) *For any $\vec{a} \in A^n$, there exists $\vec{c} \in \widehat{P}$ such that $\tilde{P}(\vec{a}) = \|\vec{a} = \vec{c}\|$.*

Proof. (1) is verified directly.

(2) Since $\widehat{P} \supseteq P$, we have $\sum_{\vec{b} \in \widehat{P}} \|\vec{a} = \vec{b}\| \geq \sum_{\vec{b} \in P} \|\vec{a} = \vec{b}\| \overset{\text{def.}}{=} \tilde{P}(\vec{a})$. On the other hand, let $\vec{a} \in A^n$, $\vec{b} \in \widehat{P}$ and $\vec{c} \in P$, we have

$$\|\vec{a} = \vec{b}\| \|\vec{b} = \vec{c}\| \leq \|\vec{a} = \vec{c}\|.$$

Summing both sides over $\vec{c} \in P$, we have $\|\vec{a} = \vec{b}\| \tilde{P}(\vec{b}) \leq \tilde{P}(\vec{a})$. Since $\vec{b} \in \widehat{P}$, we have $\tilde{P}(\vec{b}) = 1$ by the definition of $\widehat{P}$. So $\|\vec{a} = \vec{b}\| \leq \tilde{P}(\vec{a})$. This is true for any $\vec{b} \in \widehat{P}$. So $\sum_{\vec{b} \in \widehat{P}} \|\vec{a} = \vec{b}\| \leq \tilde{P}(\vec{a})$. We have thus shown that for any $(a_1, \ldots, a_n) \in A^n$, $\sum_{\vec{b} \in \widehat{P}} \|\vec{a} = \vec{b}\| = \tilde{P}(\vec{a})$. So P and $\widehat{P}$ induce the same Boolean predicate $\tilde{P}$. If $S \subseteq A^n$ also induces the Boolean predicate $\tilde{P}$ then $\tilde{P}(\vec{a}) = 1$ for $\vec{a} \in S$ and hence $S \subseteq \widehat{P}$. So, with respect to the inclusion $\subseteq$, $\widehat{P}$ is the maximal n-ary predicate of A which induces the Boolean predicate $\tilde{P}$.

(3) and (4) are done with two claims, which are special cases of main assertions.

Claim 1. $\widehat{P}$ includes the orthogonal completion of P.

Reason. Let $\vec{a}_i \in \widehat{P}$ and $\{e_i\}$, a maximal orthogonal family of B. Let $\vec{a} \in A^n$ be such that $\vec{a} \overset{\text{a.e.}}{=} \sum_i^{\perp} e_i \vec{a}_i$. Since $\widetilde{P}$ satisfies Axiom 5 by (1), we can apply Lemma 3.2. By expanding each argument and using the orthogonality of e_i, we have $\widetilde{P}(\vec{a}) = \sum_i e_i \widetilde{P}(\vec{a}_i) = \sum_i e_i = 1$. So $\vec{a} \in \widehat{P}$. This shows that $\widehat{P}$ is orthogonally complete. Since $\widehat{P} \supseteq P$, $\widehat{P}$ includes the orthogonal completion of P, as claimed.

Claim 2. Given any $\vec{a} \in A^n$, there exists $\vec{c}$ in the orthogonal completion of P such that $\widetilde{P}(\vec{a}) \geq \|\vec{a} = \vec{c}\|$.

Reason. We well-order elements of P as $\vec{b}_i$, where $i = 0, 1, \ldots$ Define $e_0 \overset{\text{def.}}{=} \|\vec{a} = \vec{b}_0\|$ and inductively for $i > 0$, $e_i \overset{\text{def.}}{=} \|\vec{a} = \vec{b}_i\| - \bigvee_{j<i} e_j$. These e_i are orthogonal to each other. Also, $\bigvee_i e_i = \bigvee_i \|\vec{a} = \vec{b}_i\| = \bigvee_{\vec{b} \in P} \|\vec{a} = \vec{b}\| \overset{\text{def.}}{=} \widetilde{P}(\vec{a})$. Set $e \overset{\text{def.}}{=} \widetilde{P}(\vec{a})$ for short. Then $1-e$ and these e_i form a maximal orthogonal subset of B. Define

$$\vec{c} \overset{\text{a.e.}}{=} \sum_i^{\perp} e_i \vec{b}_i + (1-e)\vec{b}_0.$$

Since all $\vec{b}_i \in P$, $\vec{c}$ is in the orthogonal completion of P and hence in $\widehat{P}$. By (2), $\widehat{P}$ also induces the Boolean predicate $\widehat{P}$. So we have $\widehat{P}(\vec{a}) \geq \|\vec{a} = \vec{c}\|$. On the other hand, by the definition of e_i, we have $e_i \vec{a} = e_i \vec{b}_i = e_i \vec{c}$ for all i. So

$$\|\vec{a} = \vec{c}\| \geq \bigvee_i e_i = e \overset{\text{def.}}{=} \widetilde{P}(\vec{a}).$$

So we have $\widetilde{P}(\vec{a}) = \|\vec{a} = \vec{c}\|$, where $\vec{c}$ is in the orthogonal completion of P.

Let $\vec{a} \in \widehat{P}$. Then $\widetilde{P}(\vec{a}) = 1$ by the definition of $\widehat{P}$. By Claim 2, $1 = \widetilde{P}(\vec{a}) = \|\vec{a} = \vec{c}\|$ for some $\vec{c}$ in the orthogonal completion of P. So $\vec{a}$ is in the orthogonal completion of P. This proves (3). Now, (4) follows from (3) and Claim 2. □

We come back to Beidar and Mikhalëv's theory of semiprime rings. By the orthogonal completeness of R, we have $eR \subseteq R$. Hence, for $e \in B$ and $\vec{a} = (a_1, \ldots, a_n) \in R^n$, we define $e\vec{a} \overset{\text{def.}}{=} (ea_1, \ldots, ea_n)$. We also set $\vec{0} \overset{\text{def.}}{=} (0, \ldots, 0) \in R^n$.

By (2.1), the almost equality is the same as the equality. That is, $a \overset{\text{a.e.}}{=} b$ is the same as $a = b$ for $a, b \in R$. Analogous to Lemma 3.5, we give a characterization of admissible predicates in terms of orthogonally complete predicates.

Lemma 3.9. *Assume that $B \neq \{0,1\}$. Let $\varnothing \neq P \subseteq R^n$. Then P is admissible if and only if $\vec{0} \in P$ and P is orthogonally complete.*

Proof. Assume that P is admissible. Since $P \neq \varnothing$, there exists $\vec{a} \in P$. By the admissibility of P, $\|P(\vec{a})\| = 1$. Pick $e \in B$ arbitrarily such that $0 < e < 1$. Since $0 < e \leq \|P(\vec{a})\|$, we have $e\vec{a} \in P$. So $\|P(e\vec{a})\| = 1$ again by the admissibility of P. Since $0 < 1 - e < \|P(e\vec{a})\|$, we have $\vec{0} = (1 - e)e\vec{a} \in P$, as asserted. Let $\vec{a}_i \in R^n$ and $\{e_i\}$, a maximal orthogonal subset of B. By Lemma 2.7, the Boolean predicate $\vec{a} \in R^n \to \|P(\vec{a})\| \in B$ satisfies Axiom 5. By Lemma 3.2, $\|P(\sum_i^{\perp} e_i \vec{a}_i)\| = \sum_i^{\perp} e_i \|P(\vec{a}_i)\| = 1$ since $\vec{a}_i \in P$ and hence $\|P(\vec{a}_i)\| = 1$ for all i. So $\sum_i^{\perp} e_i \vec{a}_i \in P$. This shows the orthogonal completeness of P.

On the other hand, assume that $\vec{0} \in P$ and P is orthogonally complete. Let $\tilde{P} : R^n \to B$ and $\hat{P} \subseteq R^n$ be as defined in (4) of Definition 3.7. By (1) of Lemma 3.8, $\tilde{P}$ satisfies Axiom 5 and hence is orthogonally linear by Lemma 3.4. For $\vec{a} \in R^n$ and for any $e \in B$, we have

$$\tilde{P}(e\vec{a}) = \tilde{P}(e\vec{a} + (1-e)\vec{0}) = e\tilde{P}(\vec{a}) + (1-e)\tilde{P}(\vec{0}) = e\tilde{P}(\vec{a}) + (1-e).$$

Therefore, $e\vec{a} \in \hat{P}$ iff $\tilde{P}(e\vec{a}) = 1$ iff $e = e\tilde{P}(\vec{a})$ iff $e \leq \tilde{P}(\vec{a})$. Since P is orthogonally complete, we have $P = \hat{P}$ by (3) of Lemma 3.8. So $e\vec{a} \in P$ iff $e \leq \tilde{P}(\vec{a})$. This shows the admissibility of P with $\|P(\vec{a})\| \overset{\text{def.}}{=} \|\tilde{P}(\vec{a})\|$. $\qquad\square$

For our purpose, the most important is the following characterization of predicates for which Lemma 2.12 holds.

Lemma 3.10. *Let P be a given n-ary predicate of R and $\tilde{P}$, the Boolean predicate induced by P. The following are equivalent*

(1) *P is orthogonally complete.*
(2) *For any filter F of B and for any $a_1, \ldots, a_n \in R$, $\tilde{P}_F(a_{1,F}, \ldots, a_{n,F})$ holds in the reduced product R_F iff $\overline{P}(\overline{a}_1, \ldots, \overline{a}_n)$ holds in the quotient structure R/RF'.*

Proof. Let $\hat{P} \overset{\text{def.}}{=} \{\vec{a} \in A^n : \|\tilde{P}(\vec{a})\| = 1\}$. Then $\hat{P}$ is the orthogonal completion of P by (3) of Lemma 3.8.

Assume that R_F is isomorphic to R/RF' for any filter F. Let $(a_1, \ldots, a_n) \in \hat{P}$. Then $\|\tilde{P}(a_1, \ldots, a_n)\| = 1$ by the definition of $\hat{P}$. So $\tilde{P}_F(a_{1,F}, \ldots, a_{n,F})$ holds in the reduced product R_F. Let us take $F = \{1\}$. For the filter $F = \{1\}$, the quotient structure R/RF' is merely R and $\overline{a}_i = a$ for all i. So $(a_1, \ldots, a_n) \in P$. This is true for any $(a_1, \ldots, a_n) \in \hat{P}$. So $\hat{P} \subseteq P$. But $P \subseteq \hat{P}$. So $P = \hat{P}$. We have shown that (2) implies (1).

Now, assume that $P = \widehat{P}$. Let F be a given filter of B. Assume that in the quotient structure R/RF', we have $(\bar{a}_1, \ldots, \bar{a}_n) \in \overline{P}$, where $(a_1, \ldots, a_n) \in R^n$. By the definition of the quotient structure R/RF', there exist $e \in F$ and $(c_1, \ldots, c_n) \in P$ such that $ea_i = ec_i$ for $i = 1, \ldots, n$. So

$$\tilde{P}(a_i, \ldots, a_n) \stackrel{\text{def.}}{=} \bigvee_{(b_1, \ldots, b_n) \in P} \|a_1 = b_1\| \wedge \cdots \wedge \|a_n = b_n\|$$

$$\geq \|a_1 = c_1\| \wedge \cdots \wedge \|a_n = c_n\| \geq e.$$

Since $e \in F$, $\tilde{P}_F(a_{1,F}, \ldots, a_{n,F})$ holds in the reduced product R_F. Conversely, assume that $\tilde{P}_F(a_{1,F}, \ldots, a_{n,F})$ holds in the reduced product R_F. By the definition of the reduced product R_F, $\tilde{P}(a_1, \ldots, a_n) \in F$. By (5) of Lemma 3.8, there exist $(c_1, \ldots, c_n) \in \widehat{P}$ such that

$$\tilde{P}(a_1, \ldots, a_n) = \|a_1 = c_1\| \wedge \cdots \wedge \|a_n = c_n\|.$$

So $\|a_1 = c_1\| \wedge \cdots \wedge \|a_n = c_n\| \in F$. That is, $\bar{a}_i = \bar{c}_i$ for all i. Since $(c_1, \ldots, c_n) \in \widehat{P}$ and $\widehat{P} = P$ by our assumption, $\overline{P}(\bar{a}_1, \ldots, \bar{a}_n)$ holds in the quotient structure R/RF', as asserted. $\square$

Now, we are ready to propose our generalization of Beidar and Mikhalëv's result

Theorem. *Let the semiprime ring R be an $\mathcal{L}$-structure with each function symbol designating an orthogonally linear function and with each n-ary predicate symbol designating an orthogonally complete subset of R^n. Let θ be a sentence of $\mathcal{L}$.*

(1) *Assume that θ is Horn. If θ holds on R/P for any minimal prime ideal P of R, then θ holds on any quotient ring R/eR for $e \in B$.*

(2) *Assume that $\neg\theta$ is Horn. If θ holds on any quotient ring R/eR for $e \in B$, then θ holds on R/P for any minimal prime ideal P of R.*

The proof of this theorem is verbally the same as that of Beidar and Mikhalëv's given in §2 and is therefore omitted for simplicity. Note that functions and predicates in our language $\mathcal{L}$ here may not be admissible. So the quotient ring R/eR and the subring $(1 - e)R$ may *not* be isomorphic $\mathcal{L}$-structures though they are isomorphic rings.

References

[1] K.I. Beidar and A.V. Mikhalëv, *Orthogonal completeness and algebraic systems*, Uspekhi Mat. Nauk **40**(6) (1985), 79–115, (Engl. Transl., Russian Math. Surveys **40**(6) (1985), 51–95.)

[2] J. L. Bell, Boolean Valued Models and Independence Proof in Set Theory, Clarendon Press, Oxford, 1977.

[3] C. C. Chang and H. J. Keisler, Model Theory, North-Holland Publishing Company, Amsterdam-New York-Oxford Francisco-London, 1977.

[4] T. Jech, Set Theory, Academic Press, New York-San Francisco-London, 1978.

[5] J. Lambek, Lectures on Rings and Modules, Waltham, Mass. Blaisdell Pub. Co., 1966.

[6] E. Mendelson, Introduction to Mathematical Logic, D. Van Nostrand Company, Inc., Princeton, N.J., 1964.

[7] D. S. Scott, Boolean valued models for set theory, Mimeographed notes for the 1967 American Math. Soc. Symposium on axiomatic set theory.

[8] R. M. Solovay, $2^\aleph$ *can be anything it ought to be.*, in Addison, J. W., Henkin, L. and Tarski, A. (eds.), The Theory of Models. North-Holland, Amsterdam.

Chen-Lian Chuang, Department of Mathematics, National Taiwan University, Taipei 106, Taiwan
E-mail address: chuang@math.ntu.edu.tw

Kostia Beidar and Nearrings

Yuen Fong and Wen-Fong Ke

In the fall of 1993, Kostia Beidar came to National Cheng Kung University, Tainan as a visiting scholar with the support from National Science Council, R. O. C. This visit turned into a regular position in the following year, also started a fruitful research in nearrings for Kostia besides his numerous contributions in Ring Theory and related topics.

1. Nearrings

Take an arbitrary group G, written additively but not necessarily abelian, with identity element denoted by 0. Denote by $M(G)$ the set of all mappings from G into itself. One can define an addition on $M(G)$ pointwise. For $f, g \in M(G)$, the mapping $(f + g) : G \to G$ is given by $(f + g)(x) = f(x) + g(x)$ for all $x \in G$. Then $M(G)$ is a group under this addition which is abelian only if G is. The composition of mapping is another binary operation on $M(G)$ which makes $M(G)$ a semigroup. Moreover, if $f, g, h \in M(G)$, then $(f + g) \circ h = f \circ h + g \circ h$ by definition, while $h \circ (f + g)$ and $h \circ f + h \circ g$ may not be the same as h is not a homomorphism in general. This algebraic structure, $(M(G), +, \circ)$, is one of the raw models of a nearring; more precisely, a right nearring because $\circ$ is right distributive over $+$.

Abstractly, one defines a (right) nearring to be an algebraic structure $(N, +, \cdot)$ where $(N, +)$ is a not necessarily abelian group, $(N, \cdot)$ a semigroup, and $\cdot$ is right distributive over $+$.

While the example $M(G)$ given above is a nearring which is not a ring, it is obvious that all rings are nearrings.

The theory of nearrings started when Dickson investigated the independence of the axioms of fields. He found that one side distributive law of a field does not follow from the rest of the axioms. Then he constructed many "nearfields." Some of them was used to construct "nondesarguesian" affine planes almost immediately. A small example of a nondesarguesian affine plane can be found at the end of Chapter 2 of [17].

Here we refer the readers to several books listed in the bibliography on nearrings [17, 19, 33, 37] for in depth reading of the subject.

2. The first project

Let $(G, +)$ be a group written additively which is not necessarily abelian. Let $T \subseteq G \setminus \{0\}$ and define $\cdot_T$ by $a \cdot_T b = a$, if $b \in T$; $a \cdot_T b = 0$, if $b \notin T$. Then $(G, +, \cdot_T)$ is a right *zero-symmetric* nearring, i.e. $x \cdot_T 0 = 0$ for all $x \in G$. The multiplication $\cdot_T$ is called *trivial* and so is the resulting nearring.

It is well-known that any ring whose additive group is torsion divisible has a trivial multiplication. However, it does not seem to be an easy problem to describe nearrings which have only trivial multiplications, and the problem remains open.

To tackle the problem, we investigate the structure of such a group G, if it exists. Here the basic tool used is so called $S,0$-*acts*.

Given any semigroup S with 0, a set H is called a *right $S,0$-act* if there exists a fixed element $0 \in H$ and a mapping $H \times S \to H$ with $(g, s) \mapsto gs$ such that the following conditions are satisfied:

(1) $0s = 0$ for all $s \in S$;
(2) $g0 = 0$ for all $g \in G$;
(3) $g(st) = (gs)t$ for all $s, t \in S$ and $g \in G$.

Let H and K be two right $S,0$-acts. A mapping $f : G \to H$ is said to be a *homomorphism* of the $S,0$-acts if $f(gs) = f(g)s$ for any $s \in S$ and $g \in G$.

For example, take a nearring N and the subsemigroup of right multiplications $S = \{\rho_n \mid n \in N\} \subseteq \mathrm{End}\,(N)$ (i.e., $x\rho_n = xn$ for all $x, n \in N$). Then N is a right $S,0$-act and the mapping $f : N \to \mathrm{End}\,(N)$ given by $f(n) = \rho_n$ is a homomorphism of right $S,0$-acts.

Theorem 2.1 (cf. [37, §9d]). *Let $(G, +)$ be a group and $f : G \to \mathrm{End}\,(G)$ a mapping such that (a) $S = f(G)$ is a subsemigroup of $\mathrm{End}\,(G)$, (b) $0 \in S$, and (c) $f : G \to \mathrm{End}\,(G)$ is a homomorphism of right $S,0$-acts. Define a multiplication $\cdot_f$ on G by the rule $a \cdot_f b = af(b)$ for all $a, b \in G$. Then:*

(1) *$(G, 0, +, \cdot_f)$ is a nearring;*
(2) *the multiplication $\cdot_f$ is trivial if and only if $S \subseteq \{0, id_G\}$;*
(3) *$(G, 0, +, \cdot_f)$ is a ring if and only if (i) G is an abelian group, (ii) S is a subring of the endomorphism ring $\mathrm{End}\,(G)$, and (iii) f is a homomorphism of right S-modules.*

Here we see no way of characterizing groups admits only trivial nearring multiplication. But we can use Theorem 2.1 to obtain many results about groups admitting a nontrivial nearring multiplication. Recall here that an endomorphism s of a group G is called *locally finite* if for any $a \in G$ there exist positive integers m and n such that $as^m = as^n$.

Theorem 2.2 ([12, (2.18)]). *Let $(G, +)$ be a group of order greater than 2. Suppose that G has only trivial nearring multiplications. Then:*

(1) *G is not an abelian group.*

(2) *G is not a finitely generated group.*

(3) *Any nonidentical injective endomorphism of G is a locally finite automorphism of infinite order.*

(4) *Any element of G of finite order belongs to its center $Z(G)$.*

(5) *For any $x, y \in G$ there exists a positive integer n, depending on x and y, such that $(nx) + y = y + (nx)$.*

(6) *For $x \in G$ and $m \in \mathbb{Z}$, $mx \in Z(G)$ implies that either $x \in Z(G)$ or $m = 0$.*

(7) *G is not a semidirect product of two nonzero subgroups.*

(8) *G has no nonzero nilpotent endomorphisms.*

(9) *If $s \in \mathrm{End}\,(G)$, then $G = K(s) + \mathrm{Im}(s)$.*

Sufficient conditions for a group of order greater than 2 to have no nontrivial nearring multiplications are considered as well.

Theorem 2.3. *Let G be a group of order greater than 2. Suppose that*

(a) *it has no nonzero nilpotent endomorphisms,*

(b) *it has no nontrivial idempotent endomorphisms,*

(c) *it has no nonidentical automorphisms of finite order, and*

(d) *all of its nontrivial endomorphisms are locally finite.*

Then G has no nontrivial nearring multiplications.

With this theorem, a characterization for a simple group to possess only trivial nearring multiplications can be stated:

Theorem 2.4. *Let G be a simple group with order greater than 2. Then G has no nontrivial nearring multiplications if and only if any nonidentity automorphism of G is locally finite and has an infinite order.*

The statements in the last theorem involves the "second order properties," in the language of model theory. Actually, we see that this is the only way to describe groups having no nontrivial nearring multiplications. Although the problem of characterizing groups with only trivial nearring multiplications remains open, it seems that this is a good place to end this nice project.

3. Planar Nearrings and Block Designs

Take a right nearring $(N, +, \cdot)$. For any $a, b \in N$, we say that $a \equiv b$ if $xa = xb$ for all $x \in N$. Then $\equiv$ is an equivalence on N. We say the N is *planar* if $N/\equiv$ has at least three equivalence classes, and for $a, b, c \in N$ with $a \not\equiv b$, the equation $xa = xb + c$ has a unique solution in N. Conceptually, a planar nearring is a nearring with the property that two lines of "different slops" has a unique point of intersection.

Examples of planar nearrings are abundant. For example, all finite nearfields are planar. With the following characterization of planar nearrings, one can easily construct planar nearrings from fields.

Every planar nearring gives rise to a *Ferrero pair*. Consider the group of automorphisms $\Phi = \{\rho_a \mid a \in N\}$, where $\rho_a(x) = xa$ for all $x \in N$. Then for any $\varphi \in \Phi \setminus \{1\}$, $-\varphi + 1$ is bijective. We refer to (N, Φ) as a Ferrero pair. If N is finite, this just means that Φ acts *fixed point freely* on N, i.e. if $\varphi \in \Phi \setminus \{1\}$, $x \in N$, and $\varphi(x) = x$, then $x = 0$.

Abstractly, we call a pair of groups (N, Φ) a Ferrero pair, where N is an additively written group and Φ acts on N, if for any $\varphi \in \Phi \setminus \{1\}$, $-\varphi + 1$ is bijective. Applying a standard procedure, one can define multiplications on N of a Ferrero pair (N, Φ) such that with each defined multiplication $\cdot$, N becomes a planar nearring $(N, +, \cdot)$. Moreover, the Ferrero pair corresponding to this very nearring is exactly (N, Φ). The standard procedure goes like this. Write N additively. Let S be a complete set of orbit representatives of N under the action of Φ. Take a subset $Z \subseteq S$ with $0 \notin Z$. For $a, b \in N$, define

$$ a \cdot b = \begin{cases} 0 & \text{if } b \notin \Phi(t) \text{ for all } t \in Z, \\ \varphi(a) & \text{if } b = \varphi(t) \text{ for some } t \in Z. \end{cases} $$

Then $(N, +, \cdot)$ is a planar nearring.

Taking any field F and any multiplicative subgroup P of $F \setminus \{0\}$, we have naturally a Ferrero pair (F, Φ_P) where $\Phi_P = \{\rho_a \mid a \in P\}$. Since P and Φ_P are isomorphic as groups, we usually identity Φ_P and P and say that (F, P) is a Ferrero pair. From this, we see that planar nearrings are all over the places. For example, let

$$ P = C = \{a \in \mathbb{C} \mid |a| = 1\}, $$

the unit circle. A complete set of orbit representatives of Φ_P can be taken as $S = \mathbb{R}^+ \cup \{0\}$, where $\mathbb{R}^+ = \{r \in \mathbb{R} \mid r \geq 0\}$. With $Z = \mathbb{R}^+$, one obtains the "classical" planar nearring $(\mathbb{C}, +, *)$ where $*$ is given by

$$a * b = \begin{cases} 0 & \text{if } b = 0, \\ (ab)/|b| & \text{if } b \neq 0. \end{cases}$$

Planar nearrings prove to form a very nice subset of nearrings. One main feature of planar nearrings is that they have very natural combinatorial structure called *tactical configurations* (see [17, Section 2.7.1]). In particular, every finite planar nearring also gives rise to a *balanced incomplete block design,* a combinatorial structure that draws a lot of attentions. See [17, Section 2.5] for a detailed account on this subject. Here would only describe the necessary material for our purpose.

3.1. 2-designs. Let N be a finite nearring with corresponding Ferrero pair (N, Φ). For any $a, b \in N$, we put $\Phi a + b = \{\varphi(a) + b \mid \varphi \in \Phi\}$. It holds that

(1) N is nilpotent.
(2) $\{\Phi a \mid a \in N \text{ and } a \neq 0\}$ is a partition of $N \setminus \{0\}$.
(3) $|\Phi a| = |\phi|$ for all nonzero $a \in N$.
(4) For each $x \in N$, the set $\{\Phi a + b \mid a, b \in N, a \neq 0, \text{ and } x \in \Phi a + b\}$ has exactly $|N| - 1$ members.
(5) For each pair of distinct elements $x, y \in N$, the set $\{\Phi a + b \mid a, b \in N, a \neq 0, \text{ and } x, y \in \Phi a + b\}$ has exactly $|\Phi| - 1$ members.

The properties (3), (4), and (5) make $(N, \mathbf{B}_\Phi)$, where $\mathbf{B}_\Phi = \{\Phi a + b \mid a, b \in N \text{ and } a \neq 0\}$, a balanced incomplete block design (or simply called a 2-design). Thus, members of $\mathbf{B}_\Phi$ are referred to as blocks.

Note that for any $a \in N$, the translation $T_a : N \to N$, where $T_a(x) = x + a$ for all $x \in N$, is bijective and takes a block in $\mathbf{B}_\Phi$ to a block. That is to say, the translation T_a is an automorphisms of the 2-design $(N, \mathbf{B}_\Phi)$. Because of this situation, we call $(N, \mathbf{B}_\Phi, +)$ a design group.

3.2. Circularity. The example $(\mathcal{C}, +, *)$ given above stimulates an import subclass of planar nearrings called *circular planar nearrings*. Notice that for each pair $a, b \in N$, $a \neq 0$, the set $\Phi_Z a + b = Ca + b$ is actually a circle with radius $|a|$ and center b. Any two distinct circles can at most intersect at two points, and that is where the property of being circular comes in.

Given a planar nearring N with corresponding Ferrero pair (N, Φ). It may happen that $\Phi a \cap (\Phi b + c)$ always contains no more than two elements whenever $a, b, c \in N \setminus \{0\}$. In this case, we call the nearring N, as well as the Ferrero pair (N, Φ), *circular*. Thus $(\mathcal{C}, +, *)$ is a circular planar nearring.

4. Characterization of Finite Circular Planar Nearrings

It is surprising to know that the combinatorial condition, circularity of a planar nearring, can be turned into a number theoretical and group theoretical conditions.

The first result on characterizing circular planar nearring was obtained by Modisett [35]. Let p be a prime and $k \geq 3$ an integer. Denote by $\mathcal{P}_k$ the set of prime numbers having the following property. Suppose that there is a finite field F order p^s such that k divides $p^s - 1$, and let Φ the multiplicative subgroup of F of order k. If (F, Φ) is not a circular Ferrero pair, then we put p in $\mathcal{P}_k$.

Theorem 4.1. *Given a finite field F of order p^s, p a prime, and a multiplicative subgroup Φ of F of order $k \geq 3$. Then the Ferrero pair (F, Φ) is circular if and only if $p \notin \mathcal{P}_k$. Moreover, $\mathcal{P}_k$ is a finite set for all $k \geq 3$.*

The proof of the above theorem is constructive, and based on the theory of resultants. Thus, an algorithm is readily provided for computing the sets $\mathcal{P}_k$.

Now, take an arbitrary finite Ferrero pair (N, Φ). By a result of Thompson, N is nilpotent. Therefore, N is a direct product of its Sylow p-subgroups. Moreover, let P be a Sylow p-subgroup of N. Since each P is characteristic, (P, Φ) is also a Ferrero pair. Actually, we have that (N, Φ) is circular if and only if (P, Φ) are circular for all Sylow p-subgroups P of N.

Now, analyzing deeper, one founds that for arbitrary finite p-groups, circularity again depends on p alone. This is the second project Kostia participated in nearring theory [13].

Let (N, Φ) be a finite Ferrero pair. First of all, we found that

Theorem 4.2. *If (N, Φ) is circular, the Sylow subgroups of Φ are all cyclic, i.e. Φ is metacyclic.*

Moreover, under the condition that (N, Φ) is circular, there is a finite set $\mathcal{P}_\Phi$ of primes such that

Theorem 4.3. *Let M be a finite nilpotent group. Suppose that Φ acts on M as fixed point free automorphisms. Then (M, Φ) is circular if and only if every prime divisor of $|M|$ is not contained in $\mathcal{P}_\Phi$.*

Now, the assumption that (N, Φ) is a circular Ferrero pair is crucial for the proof the above theorem. It makes sure that the set $\mathcal{P}_\Phi$ is not the whole set of primes.

To compute $\mathcal{P}_\Phi$, an algorithm is given. We use the irreducible representation of metacyclic groups as matrices over $\mathcal{C}$, and the determinant argument to obtain a set of integers. The set $\mathcal{P}_\Phi$ then consists of the prime divisors of these integers. This procedure can be carried out without the knowledge that whether there exists N

such that (N, Φ) is a circular Ferrero pair. Yet, if 0 is among these integers, then one can never find N such that (N, Φ) is circular.

Remark 4.4. The same question does not trouble Modisett's case. Because, if $k \geq 3$, then there is always a prime p with k divides $p + 1$. Take F a finite field of order p^2. Then F has a multiplicative subgroup Φ of order $p + 1$. It is known that (F, Φ) is a circular Ferrero pair. Let Ψ be the subgroup of Φ of order k. Then the Ferrero pair (F, Ψ) is also circular. Therefore, $\mathcal{P}_k$ is never the whole set of primes.

Let Φ be a metacyclic group. Then Φ admits a fixed point free action on certain groups. The question remains open under what conditions there exists a group N such that (N, Φ) is circular. (In such a case, we may say that Φ *is circular.*)

In a sequential paper [14], we give a partial solution to this question.

It is known that Φ can also be presented via generators and relations as

$$\Phi \cong \langle A, B \mid A^m = 1, B^n = A^t, BAB^{-1} = A^r \rangle = G_{m,r},$$

where m and r are two relatively prime numbers, $t = m/s$ with $s = \gcd(r-1, m)$, and n the order of r modulo m. We show that

Theorem 4.5. *Let $\Phi = G_{m,r}$ be a finite metacyclic group. Suppose that the order of r modulo m is 2, and that Φ is embeddable into the multiplicative group of some skew field. Then Φ is circular.*

We note that the assumption that Φ is embeddable in the multiplicative group of a skew field makes possible the computation of the determinants of the matrices involves. However, examples show that this is not a necessary condition.

5. Automorphism groups of certain design groups

A design $(N, \mathbf{B}_\Phi)$ arising from a finite Ferrero pair (N, Φ) has many automorphisms. They are also referred to as *multipliers*. The question of how to obtain them all is natural and hard. As we have seen in Section 3.1, every translation T_a $(a \in N)$ is an automorphism. On the other hand, if $f \in \operatorname{Aut}(N)$ normalizes Φ, i.e. $f\Phi f^{-1} = \Phi$, then if follows that

$$f(\Phi a + b) = (f\Phi f^{-1})f(a) + b = \Phi f(a) + b \quad \text{for all } a, b \in N, a \neq 0.$$

Thus, f is an automorphism of the design $(N, \mathbf{B}_\Phi)$.

It is conjectured that, except in some special situation, the translations and the normalizers of Φ are all of the automorphisms $(N, \mathbf{B}_\Phi)$ has, and the problem remains unsolved so far.

5.1. The abelian case. A restricted version of the problem is to take the group structure $(N, +)$ into consideration. That is, to find all the automorphisms of the design group $(N, \mathbf{B}_\Phi, +)$. The result is established for the cases when N and Φ are both abelian by Ke and Kiechle [31].

Theorem 5.1. *Let* (N, Φ) *be a finite Ferrero with both* N *and* Φ *are abelian. Let* $f \in \mathrm{Aut}(N)$ *be an automorphism of the design group* $(N, \mathbf{B}_\Phi, +)$. *If* $|N| > |\Phi| + 1$, *then* f *normalizes* Φ.

Note that if $|N| = k + 1$, then every permutation of N is an automorphism of the design $(N, \mathbf{B}_\Phi)$. For example, let F be the field of order 8 and let Φ be the multiplicative group of F. Then (F, Φ) is a Ferrero pair. The group $\mathrm{GL}(3, 2)$ acts on $(N, +)$ as group automorphisms; hence it acts on $(N, \mathbf{B}_\Phi, +)$ as design group automorphisms. But $\mathrm{GL}(3, 2)$ is a simple group of order 168, it cannot normalize Φ. Actually, the normalizer of Φ has only 21 elements. Thus, the condition that $|N| > |\Phi| + 1$ cannot be removed.

5.2. The general case. In early 2002, Kostia joined the discussions aiming at the general case.

Let (N, Φ) be a finite Ferrero pair, and f is an automorphism of the design group $(N, \mathbf{B}_\Phi, +)$. We say that f preserves the base blocks if

$$f(\Phi a) = \Phi f(a) \quad \text{for all } a \in N \setminus \{0\}. \tag{5.1}$$

We notice first that preserving the base blocks is a very strong condition.

(1) It can be shown that if f normalizes Φ and preserves the base blocks, then for each $\varphi \in \Phi$, there is a fixed $\lambda \in \Phi$ such that for each $x \in N \setminus \{0\}$,

$$f(\varphi(x)) = \lambda(f(x)).$$

(2) Suppose that f preserves base blocks, and that for any nonzero $s, t \in N$, there is an element $z \in N \setminus \{0\}$ (depending on s and t) such that

$$\begin{aligned} (\Phi(-s) + s) \cap (\Phi(-z) + z) &= \{0\} \quad \text{and} \\ (\Phi(-t) + t) \cap (\Phi(-z) + z) &= \{0\}. \end{aligned} \tag{5.2}$$

Then f normalizes Φ.

Here, (5.2) does not appear to be a big constraint because it holds for any N with order greater than $2k^2 - 6k + 7$, where k is the order of Φ.

Now, since N is nilpotent, and Theorem 5.1 serves as the base of an induction on the nilpotency of N, we are able to prove the following theorem (see [6]).

Theorem 5.2. *Let* (N, Φ) *be a finite Ferrero pair and let* f *be an automorphism of* $(N, \mathbf{B}_\Phi, +)$. *Put* $k = |\Phi|$. *If* $\left| N/[N, N] \right| > 2k^2 - 6k + 7$, *then* f *normalizes* Φ.

A detailed study gives us more: if $k = 3$ or $\left| N/[N, N] \right| > 2k^2 - 6k + 1$ in Theorem 5.2, then f normalizes Φ.

This is a very nice result in our opinion. However, Kostia could not see the paper been completed.

5.3. Automorphism group of other designs. A related project on computing the automorphism groups of some other 2-designs came a bit earlier than the previous one.

Clay [18] first observed that the usual Euclidean geometry of the complex plane can be obtained using a double planar nearring structure imposed on it. This was done by first factoring the multiplicative group of the complex field as the direct product of the positive reals and the unit circle, and two planar nearring structures were obtained. Then from these planar nearrings he obtained the usual rays, segments and triangles by using one of the nearrings and circles using the other. These geometric results can be extended to other fields, including finite fields, where the multiplicative group can be suitably factored.

It is natural to see some 2-designs arising from such geometrical consideration (also cf. [43]). The project started from some computation of small examples of such designs with three points, i.e. some *triple system* [26]. It is known that no description of the automorphism group of a triple system can be obtained in general. Thus, it is only possible to investigate the automorphism groups of triple systems constructed individually.

Let V be a finite dimensional vector space over a field K and let $\mathbf{L}$ be the collection of the "lines" $Ku + v, v \in V, u \in V^* = V \backslash \{0\}$. From the Fundamental Theorem of Geometry, we know that the collineations of $(V, \mathbf{L})$ are the products of the translations and the semilinear transformations of V. When K is finite, $(V, \mathbf{L})$ is a 2-design with $\lambda = 1$, and the collineations are exactly the automorphisms of the design. In this case, if we view V as a finite extension field of K, then the collineations are the maps of the form $v \mapsto \mathbf{g}(\alpha(v)) + u, v \in V$, where $\alpha : V \to V$ is a field automorphism, $\mathbf{g} \in \mathrm{GL}(V, K)$ and $u \in V$.

Now, take F a finite field of characteristic $p \neq 2$ with q elements. Take a subset $S \subseteq F$ with $|S| \geq 2$, and consider the collection

$$\mathbf{S} = \{S \cdot a + b \mid a, b \in F, a \neq 0\} = \{S^{T_{a,b}} \mid a, b \in F, a \neq 0\} = S^\Gamma,$$

where $\Gamma = \{T_{a,b} \mid a, b \in F, a \neq 0\}$, and $T_{a,b}(x) = x \cdot a + b$ for all $x \in F$. Then $(F, \mathbf{S})$ is a 2-design. In this project, we consider the case when $|S| = 3$. Since Γ

acts 2-transitively on F, there is a block $\{0, 1, t\} \in \mathbf{S}$ for some $t \in F \setminus \{0, 1\}$, and $\mathbf{S} = \{0, 1, t\}^{\Gamma}$. We thus fix such a t and denote $\mathbf{S}$ by $\mathbf{S}_t$. The cardinality of the set

$$U_t = \{t, t^{-1}, 1 - t, (1 - t)^{-1}, 1 - t^{-1}, (1 - t^{-1})^{-1}\} \tag{5.3}$$

is shown to be the parameter λ of the simple $2\text{-}(q, 3, \lambda)$ design $(F, \mathbf{S}_t)$. Explicitly, we have

$$\lambda = \begin{cases} 1, & \text{if char } F = 3 \text{ and } t = 2; \\ 2, & \text{if } t \text{ is a primitive sixth root of unity;} \\ 3, & \text{if char } F \neq 3 \text{ and } t \in \{-1, 2, \frac{1}{2}\}; \\ 6, & \text{otherwise.} \end{cases}$$

We found that unless $t \in \{-1, 2, \frac{1}{2}\}$ when char $F = 5$, or t is a root of the polynomials in P (to be given below), the set U_t has the property:

if $u, v, r, s \in U_t$ such that $ur = vs$, then $ur = 1$, or $u = v$, or $u = s$. (5.4)

This property is the key to our arguments.

Now, the set P consists of the following polynomials

$$t^2 + t - 1, \ t^2 - t - 1, \ t^2 - 3t + 1;$$
$$t^3 - t^2 - 1, \ t^3 - 2t^2 + t + 1, t^3 + t^2 - 2t + 1,$$
$$t^3 - 3t^2 + 4t - 1, \ t^3 + t - 1, \ t^3 - 4t^2 + 3t - 1; \tag{5.5}$$
$$t^4 - 3t^3 + 6t^2 - 4t + 1, \ t^4 - t + 1, \ t^4 - t^3 + 3t^2 - 3t + 1,$$
$$t^4 - 3t^3 + 3t^2 - t + 1, \ t^4 - 4t^3 + 6t^2 - 3t + 1, \ t^4 - t^3 + 1.$$

The proof of the main result is a technical lemma. It is the conditions in the lemma that leads us to consider the condition (5.4) for U_t. The lemma states

Lemma 5.3. *Let G be a finite multiplicative abelian group, M a nonempty subset of G, and $\sigma \in \text{Sym}(G)$, the permutation group on G as a set. Let $H = \langle M \rangle$, the subgroup of G generated by M. Suppose that the following conditions are satisfied:*

(i) *M contains at most one element of order 2;*
(ii) *$u^{-1} \in M$ for all $u \in M$;*
(iii) *for all $u, v \in M$ with $u \neq v$, $uM \cap vM = \{1, uv\}$;*
(iv) *$\sigma(M) = M$ and $\sigma(1) = 1$;*
(v) *there exists a map $\mu : G \to \text{Sym}(M); g \mapsto \mu_g$, such that*

$$\sigma(gu) = \sigma(g)\mu_g(u) \quad \text{for all } g \in G \text{ and } u \in M.$$

Then $\sigma(H) = H$ and $\sigma|_H$ is an automorphism of H.

In order to describe the automorphism group of $(F, \mathbf{S}_t)$, we have to use the following groups.

(1) Let $\mathcal{A}_t$ be the group of automorphisms σ of the field F such that $\sigma(U_t) = U_t$.
(2) For any subfield K of F, let $\mathrm{GL}(F, K)$ denote the group of the K-linear automorphisms of F considered as a vector space over K.
(3) Let $\mathrm{AGL}(F, K) = \{T_{\mathbf{g},b} \mid b \in F, \mathbf{g} \in \mathrm{GL}(F, K)\}$, the affine general linear group of F over K. Here $T_{\mathbf{g},b} : F \to F$ is given by $T_{\mathbf{g},b}(x) = \mathbf{g}(x) + b$ for all $x \in F$. Note that

$$T_{\mathbf{g},b} T_{\mathbf{h},d} = T_{\mathbf{gh},\mathbf{g}(d)+b} \quad \text{for all } b, d \in F, \mathbf{g}, \mathbf{h} \in \mathrm{GL}(F, K). \tag{5.6}$$

If $K = F$, then $\mathrm{GL}(F, K) \cong F^*$. In this case, we shall identify $\mathrm{GL}(F, F)$ with F^*, and so $\mathrm{AGL}(F, F) = \Gamma$.

The automorphism group of the design $(F, \mathbf{S}_t)$ is given in the following theorem.

Theorem 5.4. *Let F be a finite field of characteristic $p \neq 2$, and let $t \in F \setminus \{0, 1\}$. Let U_t be as given in (5.3), and K the subfield of F generated by U_t. Unless $t \in \{-1, 2, \frac{1}{2}\}$ and char $F = 5$, or t is a root of one of the polynomials listed in (5.5), the full automorphism group of the 2-design $(F, \mathbf{S}_t)$ is $\mathrm{AGL}(F, K)\mathcal{A}_t \subseteq \mathrm{AGL}(F, GF(p))$. Consequently, the automorphisms of $(F, \mathbf{S}_t)$ are the maps $x \mapsto \mathbf{g}(\alpha(x)) + b$, $x \in F$, where $b \in F$, $\alpha : F \to F$ is a field automorphism fixing U_t, and $\mathbf{g}$ is a linear transformation of F considered as a vector space over K.*

Here we point out the by symmetry, the result holds true when $|S| = |F| - 3$ in Theorem 5.4. We believe that the theorem could be true also for bigger subset S of F, but the proof used here does not seem to apply.

6. Maximal Right Nearring of Quotients

The third main project in the nearring theory was proposed by Kostia in 1994. Since he was so familiar with the ring of quotients, he proposed to search the possible construction of maximal right nearring of quotients for suitable nearrings.

The *maximal right ring of quotients of a prime ring with 1* was introduced by Utumi [47]. Since then, it has been an important tool in the the study of the structure of rings. The construction was later extended to semiprime rings (not necessary having 1) in [15, Chapter 2].

There are different primeness can be defined on nearrings (see [25, 48] and [5, Lemma 2.1]). We are concerned here with 3-prime and 3-semiprime nearrings.

Let N be a nearring. We say that N is 3-*prime* if for any nonzero $x, y \in N$, $xNy \neq 0$, and we say that N is 3-*semiprime* for any nonzero element $x \in N$,

$xNx \neq 0$. Also, N is said to be *equiprime* if for any $a, b, c \in N$ with $a \neq 0$, the condition $axc = axb$ for all $x \in N$ implies that $b = c$. It is easy to see that every equiprime nearring is 3-prime.

The main objects in the maximal right ring of quotients are the dense right ideals and the essential right ideals. For our construction of maximal right nearring of quotients, we found that dense right N-subsets and essential right N-subsets suitable. Here, a nonempty subset M of N is called a *right (left) N-subset* if $xN \subseteq M$ ($Nx \subseteq M$, respectively) for all $x \in M$. We remark that a left N-subset is a left N-subgroup in the sense of [37] provided that it is an additive subgroup of N. Next, a right N-subset M is said to be *dense* if for any $x, y \in N$ with $x \neq 0$ there exists some $z \in N$ such that $xz \neq 0$ and $yz \in M$. We shall use $D(N)$ to denote the set of all dense right N-subsets of N. For N-subsets I and J, set

$$\mathrm{Hom}(I_N, J_N) = \{f : I \to J \mid f(xn) = f(x)n \text{ for all } x \in N\}.$$

Note that if J is a dense N-subset and $f \in \mathrm{Hom}(N_N, J_N)$, then $f^{-1}(J)$ is also dense.

It is true that if M is a dense right N-subset if and only if it is an *essential* right N-subset, which means that $M \cap M' \neq 0$ for any nonzero right N-subset M' of N.

With these proper objects at hand, we can construct a maximal right nearring of quotients of N using $D(N)$. The construction is actually similar to that of the construction of a maximal right of quotients of a ring using dense right ideals (cf. [15, §2.1]).

Consider the set

$$\mathcal{H} = \{(f; J) \mid J \in D(N), \ f \in \mathrm{Hom}(J_N, N_N)\}.$$

Define $(f; J) \sim (g; K)$ if there exists an $L \in D(N)$ such that $L \subseteq J \cap K$ and $f = g$ on L. One can check that $\sim$ is an equivalence relation. Let $[f; J]$ denote the equivalence class determined by $(f; J) \in \mathcal{H}$. The addition and multiplication of equivalence classes in $\mathcal{H}$ are defined by

$$[f; J] + [g; K] = [f + g; J \cap K] \tag{6.1}$$

$$[f; J][g; K] = [fg; g^{-1}(J)] \tag{6.2}$$

for all $(f; J), (g; K) \in \mathcal{H}$. Then these binary operations are well-defined, and that $\mathcal{H}/\sim$ is a right zero-symmetric nearring with these operations. We shall denote the nearring just constructed by $Q_{\mathrm{mr}}(N)$.

The nearring $\mathcal{Q}$ contains N as a subnearring under the monomorphism

$$\beta : N \to \mathcal{Q}; \ x \mapsto [L_x; N],$$

where $L_x : N \to N$ is given by $L_x(a) = xa$ for all $a \in N$.

Thus for any zero-symmetric 3-semiprime nearring N, there is always a maximal right nearring of quotients $Q = Q_{mr}(N)$ which contains N as a subnearring. Moreover, this nearring Q has the following properties.

(1) Q has a unity.
(2) For all $q \in Q$ there exists a $K \in D(N)$ such that $qK \subseteq N$.
(3) For all $q \in Q$ and $J \in D(N)$, $qJ = 0$ if and only if $q = 0$.
(4) For all $J \in D(N)$ and $f : J_N \to N_N$ there exists a $q \in Q$ such that $f(x) = qx$ for all $x \in J$.

Just like the ring case, the above properties (1)–(4) characterizes $Q_{mr}(N)$. Namely, if Q' is a zero-symmetric nearring containing N as a subnearring and Q' satisfies (1)–(4), then $Q' \cong Q$.

Remark 6.1. There are also *right nearring of quotients* considered by various authors [16, 24, 29, 36, 40, 41, 42, 44]. Their constructions are analogous to the *classical right ring of quotients*. We would like to point out that the (classical) right nearrings of quotients of 3-semiprime nearrings may not exist, even if the nearrings under consideration are rings without zero divisors. On the other hand, the maximal right nearring of quotients of a 3-semiprime nearring always exists. Moreover, if a zero-symmetric 3-semiprime nearring does have a (classical) right nearring of quotients, then this right nearring of quotients is isomorphic to a subnearring of the maximal right nearring of quotients.

It is also worth mentioning that the maximal right nearring of quotients of a zero-symmetric 3-semiprime nearring N inherits some properties from N.

Theorem 6.2. *Let N be a zero-symmetric 3-semiprime nearring and let $Q = Q_{mr}(N)$, $p, q, r \in Q$, and $K \in D(N)$. Then:*

(1) *If $pKp = 0$, then $p = 0$; in particular, Q is a 3-semiprime nearring.*
(2) *If N is 3-prime and $pKq = 0$, then either $p = 0$ or $q = 0$; in particular, Q is 3-prime.*
(3) *If N is equiprime and $q \neq 0$ such that $qxp = qxr$ for all $x \in K$, then $p = r$; in particular, Q is equiprime.*

So, if N is a zero-symmetric 3-semiprime nearring and $Q = Q_{mr}(N)$, then Q is also 3-semiprime. One wonders what is the maximal right nearring of quotients, and the answer is $Q_{mr}(Q) = Q$.

When N is a 3-prime nearring with a minimal left N-subgroup, the maximal right nearring of quotients of N is a centralizer nearring determined by some group of fixed point free automorphisms.

Let L be an additive group with a group of automorphisms G acting from the right. We denote by $M_G^0(L)$ the set of all maps $f : L \to L$ such that $f(0) = 0$ and $f(x\alpha) = f(x)\alpha$ for all $x \in L$ and $\alpha \in G$. It is well-known that $M_G^0(L)$ is a right zero-symmetric nearring under pointwise addition and composition of maps, and is referred to as the *centralizer nearring on L determined by G*.

Theorem 6.3. *Let N be a zero-symmetric 3-prime nearring with minimal left N-subgroup L. Suppose that e be an idempotent in N such that $L = Ne$ is a minimal left N-subgroup of N. Let $G = eNe \setminus \{0\}$ and $\mathcal{Q} = Q_{\mathrm{mr}}(N)$. Then:*

(1) *G is a fixed point free group of automorphisms of $(L, +)$.*
(2) *$qL \subseteq L$ for all $q \in \mathcal{Q}$.*
(3) *$\mathcal{Q}$ is isomorphic to $M_G(L)$ with $\mathcal{Q}$ acting on L via left multiplication.*

The maximal right nearring of quotients of a 3-semiprime nearring N is yet to be proved useful in the study of the structure of N. We do believe that it has such potential.

7. Derivations on Transformation Nearrings

Let N be a nearring and M a subnearring of N. An additive mapping $d : M \to N$ is said to be a *derivation* of M into N if $(xy)^d = xy^d + x^d y$ for all $x, y \in M$. Here x^d denotes the image of x under d.

In what follows, let G be an additively written (not necessary abelian) group, C a fixed point free automorphism group of G, and $M_C^0(G)$ the centralizer nearring on G determined by C. For clarity in this section, we write the image of x under $\alpha \in C$ as $x\alpha$, and write the image of x under $f \in M_C^0(G)$ as $f(x)$.

Let $f \in M_C^0(G)$. We say that f is of *finite rank* if there exist finitely many elements $x_1, x_2, \ldots, x_n \in G$ such that $f(G) \subseteq \cup_{i=1}^n x_i C$, where, for each i, $x_i C = \{x_i \alpha \mid \alpha \in C\}$. When $C = \{1\}$, the nearring $M_C^0(G)$ is simply $M_0(G)$.

Let F be a skew field and V a right vector space over F. Let $R = \mathrm{End}_F(V)$ be the ring of linear transformations of the vector space V. It is well-known that if $d : R \to R$ is a derivation of the ring R, then there exists a bijective, additive transformation $T : V \to V$ such that $r^d = Tr - rT$ for all $r \in R$. Also, if $\alpha : R \to R$ is an automorphism, then there is a bijective additive transformation $S : V \to V$ such that $r^\alpha = TrT^{-1}$ for all $r \in R$ (see [30]).

In 1974, Ramakotaiah [39] proved analogous results for automorphisms of transformation nearrings (see also [37, Theorem 7.39]). The study of derivations of nearrings was initiated by Bell and Mason [4] in 1987. Since then a number of research articles on the subject have been published [1, 2, 3, 4, 21, 23, 29, 49]. Researchers mainly studied different generalizations of Posner's [38] and Herstein's [27, 28] results into the context of nearrings.

In [23, Theorem 1.1], we obtain the following result which was inspired by classical results on derivations of primitive rings with nonzero socle.

Theorem 7.1. *Let G be a nonzero additively written group, and let N be a subnearring of $M_0(G)$. If N contains all the transformations with finite rank. Then there are no nonzero derivations of N into $M_0(G)$. In particular, the transformation nearring $M_0(G)$ does not admit nonzero derivations.*

Later, Kostia joined us to continue the program, and we prove following theorem. Although the paper was finished, it was not published.

Theorem 7.2. *Let G be a nonzero additively written group with a fixed point free automorphism group C, let M be a subnearring of $M_C^0(G)$ containing all the transformations of G which are of finite rank, and let $d : M \to M_C^0(G)$ be a nonzero derivation. Then*

(1) *G is the additive group of some nearfield F, say;*
(2) *C is isomorphic to the multiplicative group $F^* = F \setminus \{0\} = G \setminus \{0\}$ that acts on G via right multiplications; and*
(3) *$M = M_C^0(G)$, and is isomorphic to F which acts on G via left multiplications.*

We note that Theorem 7.1 is a special case of Theorem 7.2 with $C = \{1\}$.

The question of whether there are nontrivial derivations on nearfields which are not fields remains open. However, using the SONATA package for GAP [45, 46], we know that there are no nontrivial derivations on the 7 exceptional finite nearfields, and this is also the case for some small Dickson nearfields provided in SONATA. On the other hand, since the set of all distributive elements of a finite nearfield is the center [37, Theorem 8.31], there are no nontrivial inner derivations on any finite nearfield.

Finally, we remark that the question of when $M_C^0(G)$ is a nearfield was considered in [32].

References

[1] Argaç, N.; Bell, H. E. Some results on derivations in near-rings, in *Nearrings and Nearfields* (Stellenbosch, 1997), pp. 42–46. Kluwer Acad. Publ., Dordrecht, the Netherlands, (2000).
[2] Beidar, K. I.; Fong Y.; Wang X. K. Posner and Herstein Theorems for Derivations of 3-prime Near-rings, *Comm. Algebra*, **24** (1996), 1581–1589.

[3] Bell, H. E. On derivations in near-rings, II, in *Near-rings, Near-fields and K-Loops* (Hamburg 1995), pp. 191–197. Kluwer Acad. Publisher, Dordrecht, the Netherlands, (1997).

[4] Bell, H. E.; Mason, G. On derivations in near-rings, in *Near-rings and Near-fields*, G. Betsch (editor). North-Holland/American Elsevier, Amsterdam, 1987.

[5] Booth, G. L.; Groenewald, N. J.; Veldsman, S. *A Kurosh-Amitsur prime radical for near-rings,* Comm. Algebra **18** (1990), 3111–3122.

[6] Beidar, K. I.; Ke, W.-F.; Kiechle, H. *Automorphisms of certain design groups II.* To appear in Journal of Algebra.

[7] Beidar, K. I.; Ke, W.-F.; Liu, C.-H.; Wu, W.-R. *Automorphism groups of certain simple 2-$(q, 3, \lambda)$ designs constructed from finite fields.* Finite Fields Appl. **9** (2003), 400–412.

[8] Beidar, K. I.; Fong, Y.; Ke, W.-F. *Maximal right nearring of quotients and semigroup generalized polynomial identity.* Results Math. **42** (2002), no. 1–2, 12–27.

[9] Fong, Y.; Ke, W.-F.; Wang, C.-S. *Nonexistence of derivations on transformation near-rings.* Comm. Algebra **28** (2000), no. 3, 1423–1428.

[10] Beidar, K. I.; Fong, Y.; Ke, W.-F.; Wu, W.-R. *On semi-endomorphisms of groups.* Comm. Algebra **27** (1999), no. 5, 2193–2205.

[11] Beidar, K. I.; Fong, Y.; Ke, W.-F. *On the simplicity of centralizer nearrings.* First International Tainan-Moscow Algebra Workshop (Tainan, 1994), pp. 139–146, de Gruyter, Berlin, 1996.

[12] Beidar, K.; Fong, Y.; Ke, W.-F.; Liang, S.-Y. *Nearring multiplications on groups.* Comm. Algebra **23** (1995), no. 3, 999–1015.

[13] Beidar, K.; Fong, Y.; Ke, W.-F. *On finite circular planar nearrings.* J. Algebra **185** (1996), no. 3, 688–709.

[14] Beidar, K. I.; Ke, W.-F.; Kiechle, H. *Circularity of finite groups without fixed points.* Monatsh. Math. **144** (2005), no. 4, 265–273.

[15] Beidar, K. I.; Martindale, W. S., 3rd; Mikhalev, A. V. Rings with Generalized Identities, Marcel Dekker, Inc., 1996.

[16] Chan, G. H.; Chew, K. L. *On extensions of near-rings,* Nanta Math. **5** (1971), 12–21.

[17] Clay, J. R. Nearrings. Geneses and applications. Oxford Science Publications. The Clarendon Press, Oxford University Press, New York, 1992.

[18] Clay, J. R. *Geometry in fields.* Algebra Colloq. **1** (1994), 289–306.

[19] Cotti Ferrero, C.; Ferrero, G. Nearrings. Some developments linked to semigroups and groups. Advances in Mathematics (Dordrecht), 4. Kluwer Academic Publishers, Dordrecht, 2002.

[20] Dembowski, P. Finite geometries. Ergebnisse der Mathematik und ihrer Grenzgebiete, Band 44 Springer-Verlag, Berlin-New York 1968

[21] Dheena, P.; Rajeswari, C. *On near-rings with derivations,* J. Indian Math. Soc., **60** (1994), 267–271.

[22] Ferrero, G. *Sui moltiplicatori (nel senso di Hall), e sui disegni ricchi di moltiplicatori,* (Italian) Atti del Convegno di Geometria Combinatoria e sue Applicazioni (Univ. Perugia, Perugia, 1970), pp. 233–237. Ist. Mat., Univ. Perugia, Perugia, 1971.

[23] Fong, Y.; Ke, W.-F.; Wang, C. S. Nonexistence of Derivations on Transformation Near-rings, *Comm. Algebra,* **28** (2000), 1423–1428.

[24] Graves, J. A.; Malone, J. J. *Embedding near domains,* Bull. Austral. Math. Soc. **9** (1973), 33–42.

[25] Groenewald, N. J. *Different prime ideals in nearrings,* Comm. Algebra **19** (1991), 2667–2675.

[26] Hall, Marshall, Jr. Combinatorial theory. Blaisdell Publishing Co. Ginn and Co., Waltham, Mass.-Toronto, Ont.-London 1967.

[27] Herstein, I. N. *A note on derivations,* Canad. Math. Bull., **21** (1978), 369–370.

[28] Herstein, I. N. *A note on derivations II,* Canad. Math. Bull., **22** (1979), 509–511.

[29] Holcombe, Wm. M. L. *Near-rings of quotients of endomorphism near-rings,* Proc. Edinburgh Math. Soc. **19** (1974/75), 345–352.

[30] Jacobson, N. *Structure of Rings.* Amer. Math. Soc. Colloquium Publ., Providence, 1964.

[31] Ke, W.-F.; Kiechle, H. *Automorphisms of certain design groups.* J. Algebra **167** (1994), no. 2, 488–500.

[32] Maxson, C. J.; Smith, K. C. *The centralizer of a set of group automorphisms,* Comm. Algebra **8** (1980), 211–230.

[33] Meldrum, J. D. P. Near-rings and their links with groups. Research Notes in Mathematics, 134. Pitman (Advanced Publishing Program), Boston, MA, 1985.

[34] Modisett, M. C. *A characterization of the circularity of certain balanced incomplete block designs.* Ph. D. dissertation, University of Arizona, 1988.

[35] Modisett, M. C. *A characterization of the circularity of balanced incomplete block designs.* Utilitas Math. **35** (1989), 83–94.

[36] Oswald, A. *Near-rings of quotients,* Proc. Edinburgh Math. Soc. **22** (1979), 77–86.

[37] Pilz, G. Near-rings. The theory and its applications. Second edition. North-Holland Mathematics Studies, 23. North-Holland Publishing Co., Amsterdam, 1983.

[38] Posner, E. *Derivations in prime rings,* Proc. Amer. Math. Soc., **8** (1957), 1093–1100.

[39] Ramakotaiah, D. *Isomorphisms of near-rings of transformations,* J. London Math. Soc. **9** (1974), 272–278.

[40] Seth, V. Near-rings of quotients, Doctoral Dissertation, Indian Institute of Technology, 1974.

[41] Seth, V.; Tewari, K. *Classical near-rings of left and right quotients,* Prog. Math. **12** (1978), 115–123.

[42] Shafi, M. *A note on quotient near-ring,* Arabian J. Sci. Eng. **4** (1979), 59–62.

[43] Sun, H.-M. Planar Nearrings and Block Designs. Ph. D. Thesis, University of Arizona, Tucson, 1995.

[44] Tewari, K. *Quotient near-rings and near-rings modules,* Oberwolfach, 1972.

[45] The GAP Group. GAP—Groups, Algorithms, and Programming, Version 4.2, Aachen, St. Andrews, 2000. (http://www-gap.dcs.st-and.ac.uk/~gap)

[46] The SONATA Team. SONATA—Systems Of Nearrings And Their Applications, Version 1, 1997. Institut für Algebra, Stochastik und wissensbasierte mathematische Systeme, University of Linz, Austria.

[47] Utumi, Y. *On quotient rings,* Osaka J. Math. **8** (1956), 1–18.

[48] Veldsman, S. *On equiprime nearrings,* Comm. Algebra **20** (1992), 2569–2587.

[49] Wang, X. K. *Derivations in prime near-rings*, Proc. Amer. Math. Soc. **121** (1994), 361–366.

Yuen Fong, Department of Mathematics, National Cheng Kung University, Tainan 701, Taiwan
E-mail address: fong@mail.ncku.edu.tw

Wen-Fong Ke, Department of Mathematics, National Cheng Kung University, Tainan 701, Taiwan
E-mail address: wfke@mail.ncku.edu.tw

Kostia Beidar's Contributions to Module and Ring Theory

Christian Lomp and Robert Wisbauer

Abstract. At the beginning of his mathematical career, Kostia Beidar was work-ing on rings with polynomial identities and primeness conditions for rings. By Posner's theorem, the two-sided quotient ring of a prime PI-ring is a finite ma-trix ring over some field. This result was extended by Martindale to rings with generalized polynomial identities by the construction of the central closure of a prime ring. Kostia was working extensively in this setting and made crucial contributions to the understanding of the theory. While his contribution to gen-eral PI theory will be outlined elsewhere we want to sketch here his work on prime rings and the resulting study of (strongly) prime modules. An account on his papers on Hopf algebras is given and attention is drawn to some more recent constructions which grew out from Kostia's basic contributions to this field.

Contents

1. Rings and rings of quotients

1.1. Finite automorphism groups. The two early papers [6] and [5] by Kostia are dealing with automorphism groups of algebras. Let G be a finite group acting on an associative ring A with unity, with A^G the fixed ring of this action. An

The first author was partially supported by the Centro de Matemática da Universidade do Porto, financed by FCT (Portugal) through the programmes POCTI and POSI, with national and European Community structural funds.

element $\gamma \in A$ is called an *element of trace* one if

$$\mathrm{tr}_G(\gamma) = \sum_{g \in G} g(\gamma) = 1.$$

In [6] it is proved that if A has an element of trace one, then A inherits each of the following two properties from A^G:

(i) every quotient ring by a primitive ideal is Artinian;

(ii) the ring is a PI-ring.

The second property was shown later by Kostia and B. Torrecillas in [44] to hold also for finite dimensional Hopf algebras H with cocommutative coradical acting on a unital algebra A such that $t \cdot \gamma = 1$ for some $\gamma \in A$ and a left integral $t \in H$.

Denote by $N(A)$ the upper nil-radical of the algebra A. Herstein's conjecture asks whether $A/N(A)$ is commutative, provided that A^G lies in the centre of A and G is a cyclic group of prime order. Kostia gave an affirmative answer in [6] under the additional assumption that A^G is semiprime or that A has an element $\gamma \in A$ such that $\mathrm{tr}_G(\gamma)$ is central in A and not a zero divisor of A.

Under the global assumption that G is a finite cyclic group and A^G lies in the centre of A, he had shown already in [5, Theorem 4] that the existence of an element $\gamma \in A$ where $\mathrm{tr}_G(\gamma)$ is central and not a zero divisor in A implies the commutativity of $A/\mathrm{rad}\,A$, where $\mathrm{rad}\,A$ is the classical radical of A. Moreover [5, Theorem 2] shows that the action of G on A can be extended to an action of G on $Q(A)$, the maximal right ring of quotients of A, provided A contains a central element of trace one and satisfies $\mathrm{rad}\,A = 0$. In this situation one also has $Q(A)^G = Q(A^G)$.

For convenience we recall that the *maximal (or complete) right ring of quotients* of a ring A is defined as the ring

$$Q(A) = \{b \in E(A) \mid \text{for any } f \in \mathrm{End}\,E(A), f(A) = 0 \text{ implies } f(b) = 0\},$$

where $E(A)$ denotes the injective hull of A as right A-module.

1.2. Non-degenerated alternative rings. Kostia has written several joint papers on nonassociative rings. Recall that a ring A is said to be *alternative* if $x^2 y = x(xy)$ and $yx^2 = (yx)x$ for all $x, y \in A$. An alternative ring A is *non-degenerate* if $xAx \neq 0$ for any non-zero element $x \in A$. In [20] it is proved:

Theorem. *For a non-degenerate alternative ring A, the following are equivalent:*

(a) *A is prime, that is $IJ \neq 0$ for any two nonzero ideals $I, J \subset A$;*

(b) $(aA)b \neq 0$ *for any nonzero* $a, b \in A$;

(c) $a(Ab) \neq 0$ *for any nonzero* $a, b \in A$.

A similar result is obtained for Jordan rings in [20, Theorem 2] where non-degeneracy is defined by the Jordan triple product of elements.

The study of non-degenerate alternative rings was continued in [21]. There the construction of *nearly classical localization* is given and the structure of non-degenerate alternative algebras is described. Purely alternative (that is, not associative) alternative rings are called *generalized Cayley-Dickson rings*.

Theorem ([21, Corollary 2.15]). *Let A be a non-degenerate alternative algebra. Then A is a subdirect product of semiprime associative algebras and of generalized Cayley-Dickson rings.*

That Cayley-Dickson rings play a dominant role for purely alternative rings was shown in [21, Theorem 2.16]:

Theorem. *Let A be a non-degenerate alternative algebra over a commutative associative noetherian ring R with unit. Then A is either an associative algebra or A contains a subalgebra which is a Cayley-Dickson ring.*

In Kostia's paper [30] it is proved that the centre of a non-degenerate purely alternative algebra A contains a dense ideal I such that for any nonzero $t \in I$ the classical localization A_t of the algebra A with respect to t is a Cayley-Dickson algebra over its centre. This is used to show that the classical ring of quotients of an alternative PI-algebra is a PI-algebra and some of the results are applied to the description of von Neumann regular alternative algebras. Let us mention that purely alternative prime algebras behave similar to associative prime PI-algebras. For example, for both types of rings the nonzero ideals have nonzero intersections with the centre.

1.3. Orthogonal completeness. Although it is generally useful to study semi-prime rings by reducing the questions to prime rings, this approach frequently presents some difficulties. For example, it is well known that every polynomial identity of a prime ring R is also an identity of its maximal right ring of quotients $Q(R)$. However, when trying to prove a similar statement for semiprime rings, the direct reduction to prime rings is not so easy, since in general there is no homomorphism $Q(R) \to Q(R/P)$, where P is a prime ideal of the ring R.

Some of the problems arising in this context can be overcome by the theory of *orthogonal completeness* which was investigated and developed in a series of papers by Kostia mainly in cooperation with A. V. Mikhalev [16, 18, 19, 17] and a nice overview of these results is given in [37].

Recall that a unital ring B is called *Boolean* if every element of B is an idempotent. An algebra A over a Boolean ring B is said to be *orthogonally complete* if for any $a \in A$ the ideal

$$r(B; a) = \{ b \in B \mid ab = 0 \}$$

is principal, and if for any dense orthogonal subset $E \subseteq B$ and for any family of elements $S = \{ s_e \mid e \in E \} \subseteq A$, there exists an element $a \in A$ such that $ea = es_e$ for all $e \in E$.

One way of obtaining an orthogonal completion of an algebra A over an associative semiprime commutative ring K is by *almost classical localization*. For this, let $\mathscr{F}$ be the filter of dense ideals of K, and assume that A is $\mathscr{F}$-torsion free. It turns out that $A_{\mathscr{F}}$ is an orthogonally complete $K_{\mathscr{F}}$-algebra over the Boolean ring B of idempotents of $K_{\mathscr{F}}$. The orthogonal completion of A is then the intersection of all orthogonally complete subalgebras of $A_{\mathscr{F}}$ containing A.

The method of orthogonal completeness has three components:

(1) Constructions and descriptions of orthogonal completions,

(2) sufficient conditions for the primeness of Pierce stalks of orthogonally complete rings, and

(3) metatheorems which allow one to lift structure theorems to the orthogonally complete rings from their Pierce stalks.

The theory was applied by the authors in various situations, for example to study the structure of nondegenerate alternative algebras and the structure of semiprime rings with bounded indices of nilpotent elements. We mention two typical results, [9, Theorem 2 and 3].

Theorem. *Let $Q = Q(A)$ be the maximal right ring of quotients of the prime ring A. Then the following conditions are equivalent:*

(a) *A is a right Goldie ring;*

(b) *there exists some $n \in \mathbb{N}$ such that $a^n \in a^{n+1}Q$ for all $a \in A$;*

(c) *the indices of the nilpotent elements of A are bounded, and for each $a \in A$ there exists $n = n(a) \in \mathbb{N}$ such that $a^n \in a^{n+1}Q$.*

Theorem. *Let $Q = Q(A)$ be the maximal right ring of quotients of the semiprime ring A. Then the following are equivalent:*

(a) *Q is the direct sum of finitely many matrix rings over strongly regular right self-injective rings;*

(b) *there exists some $n \in \mathbb{N}$ such that $a^n \in a^{n+1}Q$ for all $a \in A$.*

1.4. The central closure of a prime ring. Let R be prime ring and let $\mathbb{U} = \{U \mid U$ is a nonzero ideal in $R\}$. For any non-zero $U, V \in \mathbb{U}$, consider homomorphisms $f : U_R \to R_R$ and $g : V_R \to R_R$ and define f to be equivalent to g if $f|_{U \cap V} = g|_{U \cap V}$. This defines an equivalence relations on the set of all such morphisms. With obvious addition and multiplication the set of the equivalence classes form a ring Q, the *Martindale ring of quotients* of A. The centre C of Q is a field and is called the *extended centroid* of R. Then $S = RC \subset Q$ is a prime ring which is called the *central closure* of R.

In [56, Theorem 3] it is shown that S satisfies a generalized polynomial identity if and only if S contains a minimal right ideal eS, $e^2 = e \in S$, and eSe is a finite dimensional division algebra over C.

Let R be a semiprime ring. As outlined in [57], the above construction can be repeated by replacing the set $\mathbb{U}$ of all ideals by all essential ideals of R. It was shown by Kharchenko that in this case the extended centroid C is a regular ring and Kostia proved (in [1, Theorem 1]) that C is a self-injective ring. Moreover the central closure $S = RC$ is a semiprime ring. As shown in [52] associativity of the ring R is not needed for the construction of the central closure. This will also follow from the module theoretic constructions considered in 3.2.

When is the central closure simple? One of the questions of central localization is when the central closure $S = RC$ is a simple ring. If R be a prime ring, then S is a prime ring with the centre C (extended centroid) being a field. If S satisfies a polynomial identity, then the intersection of any ideal with the centre C is non-zero and hence contains an invertible element. Thus S is a simple ring. Moreover, PI theory tells us that S has finite dimension as C-vector space.

The question arises which property of the ring R (other than PI) implies that S is a simple ring. This may also be expressed by properties of the (R, R)-bimodule R and this will be done in 3.1. To prepare this some new notions in module theory are needed.

2. Strongly prime and semiprime modules

To provide the techniques to deal with the questions asked above recall that for any R-module M, the full subcategory of R-Mod whose objects are submodules of M-generated modules is denoted by $\sigma[M]$. This is a Grothendieck category and every object has an (M-)injective hull in $\sigma[M]$.

Let $\widehat{M}$ denote the M-injective hull of $M \in \sigma[M]$. The class

$$\{X \in \sigma[M] \mid \operatorname{Hom}_R(X, \widehat{M}) = 0\}$$

is a torsion class and induces a torsion theory in $\sigma[M]$.

We write $U \trianglelefteq M$ to indicate that U is an essential submodule of M. The module M is called *polyform* provided $\operatorname{Hom}_R(M/U, \widehat{M}) = 0$ for every $U \trianglelefteq M$. Notice that a ring R is left polyform if and only if its left singular submodule is zero.

2.1. Bimodule properties of polyform modules. Of particular interest is the bimodule structure of polyform modules. Motivated by the properties observed for nonsingular and semiprime rings and Kostia's experience with idempotents, the following is shown in [34, 3.3].

Theorem. *Let M be a polyform R-module, $\widehat{M}$ its M-injective hull and $T = \operatorname{End}_R(\widehat{M})$. Denote by C the center of T (i.e., the endomorphism ring of $\widehat{M}$ as an (R, T)-bimodule). Then:*

(1) *Every essential (R, T)-submodule of $\widehat{M}$ is essential as an R-submodule.*

(2) *$\widehat{M}$ is self-injective and polyform as an (R, T)-bimodule.*
    C is a regular self-injective ring.

(3) *For every submodule (subset) $K \subset \widehat{M}$, there exists an idempotent $\varepsilon(K) \in C$, such that $\operatorname{An}_C(K) = (1 - \varepsilon(K))C$.*

(4) *If $K \trianglelefteq L \subset \widehat{M}$, then $\varepsilon(K) = \varepsilon(L)$.*

(5) *Every finitely generated C-submodule of $\widehat{M}$ is C-injective.*

(6) *If $\widehat{M}$ is a finitely generated (R, T)-module, $\widehat{M}$ is a generator in C-Mod.*

These observations lead to the definition and properties of the

Theorem (Idempotent closure of polyform modules). *Let M be an R-module, $T = \operatorname{End}_R(\widehat{M})$ and B the Boolean ring of all central idempotents of T. Let $\widetilde{M} = MB$. Then for every $a \in \widetilde{M}$, there exist $m_1, \ldots, m_k \in M$ and pairwise orthogonal $c_1, \ldots, c_k \in B$ such that $a = \sum_{i=1}^{k} m_i c_i$. If M is a polyform module, there exist pairwise orthogonal idempotents $e_1, \ldots, e_k \in B$ such that*

(1) *$a = \sum_{i=1}^{k} m_i e_i$;*

(2) *$e_i = \varepsilon(m_i) e_i \quad for \; i = 1, \ldots, k$;*

(3) *$\varepsilon(a) = \sum_{i=1}^{k} e_i$.*

2.2. Strongly semiprime modules. Again this point of view can be extended to semiprimeness and here Kostia came in with substantial contributions in [34], [35], and [36]. A module M is called *strongly semiprime*, for short SSP, if its M-injective hull $\widehat{M}$ is semisimple as an (R,T)-bimodule where $T = \text{End}_R(\widehat{M})$ (see [35, 4.5]).

Theorem. *Let M be an R-module.*

   (1) *Assume M has an essential socle and for every $N \trianglelefteq M$, $M \in \sigma[N]$. Then M is semisimple.*

   (2) *M is semisimple if and only if every module in $\sigma[M]$ is SSP.*

In general SSP modules need not be polyform. However we have the following theorem:

Theorem (Projective strongly semiprime modules). *Let M be projective in $\sigma[M]$ and $T = \text{End}_R(\widehat{M})$. Then the following are equivalent:*

   (a) *M is an SSP-module;*

   (b) *M is polyform and for any $N \trianglelefteq M$, $M \in \sigma[N]$.*

Applied to the case $M = R$ we get a characterization of left SSP rings ([35, 8.2]):

Theorem. *For a ring R let $Q = Q(R)$ denote the maximal left ring of quotients. Then the following are equivalent:*

   (a) *R is left SSP;*

   (b) *for every essential left ideal $N \subset R$, $R \in \sigma[N]$;*

   (c) *every $N \trianglelefteq {}_RR$ contains a finite subset X with $An_R(X) = 0$;*

   (d) *R is semiprime and every left ideal $I \subset R$ contains a finite subset $X \subset I$ with $An_R(X) = An_R(I)$;*

   (e) *Q is a semisimple (R, Q)-module.*

If R satisfies these conditions, then Q is left self-injective, von Neumann regular, and a finite product of simple rings. Left ideals with property (c) in 2.2 are also called *insulated*. So the rings described here are exactly the *left strongly semiprime rings* as considered by Handelman [51].

2.3. Strongly prime modules. The R-module M is called *strongly prime* if every nonzero submodule $K \subset M$ is a subgenerator in $\sigma[M]$, that is, $M \in \sigma[K]$.

Theorem. *For an R-module M with $T = \text{End}_R(\widehat{M})$, the following are equivalent:*

(a) M *is strongly prime;*

(b) M *is SSP and* $\widehat{M}$ *is a uniform* (R,T)*-bimodule;*

(c) $\widehat{M}$ *is a simple* (R,T)*-bimodule;*

(d) $\widehat{M}$ *has no fully invariant submodule.*

In particular, for a uniform R-module M, the conditions strongly prime and SSP are equivalent.

3. The bimodule structure of an algebra

As pointed out earlier, the motivation for some of the notions introduced for modules was to understand the bimodule structure of an algebra. This will be outlined in this section.

3.1. Bimodule structure of an algebra. For any algebra (or ring) A and $a \in A$, the left and right multiplications

$$L_a : A \to A, \ x \mapsto ax, \quad R_a : A \to A, \ x \mapsto xa,$$

are $\mathbb{Z}$-linear maps and the *multiplication algebra* $M(A)$ of A is defined as the subring of $\mathrm{End}_{\mathbb{Z}}(A)$ generated by all L_a, R_a, $a \in A$ and the identity map of A. Notice that we do not require A to be associative nor to have a unit.

Then A is a left module over $M(A)$ and $\mathrm{End}_{M(A)}(A)$ is called the *centroid* of A. If A has a unit, then this is isomorphic to the center $Z(A)$ of A.

In general A is not a generator in $M(A)$-Mod. To relate properties of A with properties of $M(A)$-modules one has to restrict to the full subcategory $\sigma[A]$ of $M(A)$-Mod whose elements N are subgenerated by A, that is, N is a submodule of an A-generated $M(A)$-module. If A has a unit, then an $M(A)$-module N is A-generated if and only if it is generated by its central elements $\{m \in N \mid am = ma \text{ for all } a \in A\}$.

Notice that in the category $\sigma[A]$, every object has an injective hull. In particular, the selfinjective hull $\widehat{A}$ of the $M(A)$-module A is injective in $\sigma[A]$ and is an A-generated $M(A)$-module, that is, $\widehat{A} = A\mathrm{Hom}_{M(A)}(A, \widehat{A})$.

3.2. Central closure of semiprime algebras. For the construction of the maximal left ring of quotients of a semiprime associative ring it is of interest if the ring is left non-singular. Although every semiprime commutative ring is non-singular, a noncommutative semiprime ring need not be non-singular as a left (or right) module over itself. However, any semiprime ring A is non-singular in the category

$\sigma[A]$ as an (A, A)-bimodule and this makes it possible to construct a quotient ring for any semiprime ring.

Theorem ([35, 9.1]). *Let A be a semiprime algebra with A-injective hull $\widehat{A}$ as $M(A)$-module and $T := \mathrm{End}_{M(A)}(\widehat{A})$ (the extended centroid). Then:*

(1) *A is a polyform $M(A)$-module.*

(2) *T is a commutative, regular, and self-injective ring.*

(3) *$\widehat{A} = AT$ is a semiprime ring with respect to the multiplication*

$$(as) \cdot (bt) := (ab)st, \quad for \ a, b \in A, \ s, t \in T,$$

and linear extension.

$\widehat{A}$ is called the *central closure* of A.

In the given situation the idempotent closure of a polyform module yields a ring extension.

Theorem (Idempotent closure of semiprime algebras). *Let A be a semiprime R-algebra, $T = \mathrm{End}_{M(A)}(\widehat{A})$, B the Boolean ring of idempotents of T. The idempotent closure of A as an $M(A)$-module, $\widetilde{A} = AB$ (see 2.1), is an R-algebra and*

(1) *for any $a \in \widetilde{A}$, there exist $a_1, \ldots, a_k \in A$ and pairwise orthogonal elements $e_1, \ldots, e_k \in B$, such that*

    (i) *$a = \sum_{i=1}^{k} a_i e_i$,*

    (ii) *$e_i = \varepsilon(a_i)e_i$, for $i = 1, \ldots, k$, and*

    (iii) *$\varepsilon(a) = \sum_{i=1}^{k} e_i$.*

(2) *For every prime ideal $K \subset \widetilde{A}$, $P = K \cap A$ is a prime ideal in A and*

$$\widetilde{A}/K = (A + K)/K \simeq A/P.$$

The set $x = \{e \in B | \widetilde{A}e \subset K\}$ is a maximal ideal in B and $K = PB + \widetilde{A}x$.

(3) *For any prime ideal $P \subset A$, there exists a prime ideal $K \subset \widetilde{A}$ with $K \cap A = P$.*

Of course for any prime algebra the central closure can be constructed as in 3.2 and we obtain special properties.

Theorem (Central closure or prime algebras). *Let A be a prime algebra with A-injective hull $\widehat{A}$ as $M(A)$-module and $T := \mathrm{End}_{M(A)}(\widehat{A})$. Then:*

(1) *T is a field.*

(2) $\widehat{A} = AT$ is a prime ring whose center is a field.

(3) $\widehat{A}$ is a simple ring if and only if A is strongly prime as an $M(A)$-module.

Recall that a ring A is an *Azumaya ring* if A is a generator in the category $\sigma[A]$ of $M(A)$-modules. We mention the result [34, 9.9] showing some relations with this type of algebras.

Theorem. *Let A be a semiprime ring with unit, $T = \text{End}_{M(A)}(\widehat{A})$ where $\widehat{A}$ is the central closure of A (see 3.2). Then the following are equivalent:*

(a) $\widehat{A}$ *is an Azumaya ring;*

(b) $\widehat{A}$ *is a biregular ring and $\widehat{A}$ is a projective module in $\sigma_{M(\widehat{A})}[\widehat{A}]$;*

(c) *the $M(\widehat{A})$-module $\widehat{A}$ is a generator in $\sigma_{M(\widehat{A})}[\widehat{A}]$;*

(d) $M(\widehat{A})$ *is a dense subring in $\text{End}_T(\widehat{A})$.*

3.3. Central closure for Hopf module algebras. Given an algebra A with an action of a Hopf algebra H, we can similarly construct a kind of central closure for A to which we can extend the action of H. More generally, the above construction works for any extension $A \subseteq B$ of unital rings such that there exists a ring homomorphism $\varphi : B \to \text{End}(A)$ with $M(A) \subseteq \text{Im}(\varphi)$. By [55] we have the following theorem:

Theorem. *Let $A \subseteq B$ as above and denote by $\widehat{A}$ the self-injective hull of A as a B-module. Assume that A is B-semiprime, i.e., has no non-zero B-stable nilpotent ideal. Then*

(1) *A is a polyform B-module and $A^B = \text{End}_B(A)$ is a commutative reduced ring.*

(2) *$\widehat{A}^B = \text{End}_B(\widehat{A})$ is a commutative von Neumann regular self-injective ring, which is a field if and only if A is B-prime, i.e. the product of two non-zero B-stable ideals of A is non-zero.*

(3) *$\widehat{A} = A\widehat{A}^B$ is a semiprime ring with respect to*

$$(as) \cdot (bt) := (ab)st, \quad \text{for } a, b \in A, \ s, t \in \widehat{A}^B,$$

and linear extension.

Since $\widehat{A}$ is a B-module, the action of B on A extends to an action of B on $\widehat{A}$. As an application to Hopf algebra action one defines a new multiplication on the tensor product $B = A^e \otimes H$, where $A^e = A \otimes A^{op}$ is the enveloping algebra and H is a Hopf algebra acting on A, such that $A \subseteq B$ is an extension as described above. The construction of $\widehat{A}$ yields a new H-module algebra which coincides with the central closure constructed by Matczuk.

3.4. Strongly semiprime algebras. Under some non-degeneracy condition strongly semiprime algebras are semiprime and thus the central closure is defined yielding [35, 9.4].

Theorem. *Let A be a ring which is not annihilated by any non-zero ideal and $T = \mathrm{End}_{M(A)}(\widehat{A})$. Then the following conditions are equivalent:*

(a) *A is an SSP $M(A)$-module;*

(b) *A is a semiprime algebra and for every essential ideal $U \subset A, A \in \sigma_{M(A)}[U]$;*

(c) *A is semiprime and the central closure $\widehat{A}$ is a direct sum of simple ideals.*

If A is associative, then (a)–(c) are equivalent to:

(d) *A is semiprime and for every ideal $I \subset A$, $A/\mathrm{An}_A(I) \in \sigma_{M(A)}[I]$.*

4. Hopf algebras and quantum Yang-Baxter equation

In two papers Kostia, in cooperation with A. Stolin and Y. Fong, proved a couple of interesting results on Frobenius algebras over commutative rings that they could apply successfully to Hopf algebras over commutative rings.

4.1. Frobenius algebras and quantum Yang-Baxter equation. An algebra A over a commutative ring K is called *Frobenius* if it is a finitely generated projective K-module and there exists $\phi \in A^* = \mathrm{Hom}(A, K)$ such that the map $\psi : A \to A^*$ with $\psi(x)(y) = \phi(yx)$ for all $x, y \in A$ is an isomorphism of K-modules. Since A is finitely generated and projective as K-module, it has a dual basis $\{e_i, f^i\}_{1 \le i \le n}$, i.e. $e_i \in A$ and $f^i \in A^*$ such that

$$x = \sum_{i=1}^{n} f^i(x)e_i.$$

Then given any $x \in A$, there exists a unique $x' \in A$ such that

$$\phi(yx') = \psi(x')(y) = \phi(xy)$$

for all $y \in A$. This defines an automorphism $\alpha : A \to A$ with $\alpha(x) = x'$, called the *Nakayama automorphism* of A.

Let $Q = \sum a_i \otimes b_i \in A \otimes A$ be an element of the tensor product. We will use the following notations for elements in the 3-fold tensor of A:

$$Q^{12} = \sum a_i \otimes b_i \otimes 1 \in A^{\otimes 3},$$

$$Q^{13} = \sum a_i \otimes 1 \otimes b_i \in A^{\otimes 3},$$

$$Q^{23} = \sum 1 \otimes a_i \otimes b_i \in A^{\otimes 3}.$$

If $R \in \mathrm{End}_K(A \otimes A)$ then we also use the notation R^{12}, R^{13}, R^{23} to denote the endomorphisms of the 3-fold tensor $A^{\otimes 3}$ acting on the components indicated by the superscripts, i.e. $R^{12} = R \otimes id$, $R^{23} = id \otimes R$ and $R^{13}(x \otimes y \otimes z) = (id \otimes T)R(x \otimes z) \otimes y$ where T is the twist map $T(a \otimes b) = b \otimes a$.

Theorem ([39, Theorem 3.4]). *Let A be a Frobenius algebra over K with Frobenius homomorphism ϕ, Nakayama automorphism α, and dual basis (e_i, f^i) and $e^i := \psi^{-1}(f^i)$. Set $Q = \sum_{i=1}^n e_i \otimes e^i \in A \otimes_K A$ and define $T \in \mathrm{End}_K(A \otimes_K A)$ by $T(a \otimes b) = b \otimes a$. Then Q satisfies the braid relation*

$$Q^{12}Q^{23}Q^{12} = Q^{23}Q^{12}Q^{23}$$

and $R = QT \in \mathrm{End}_K(A \otimes_K A)$ satisfies the quantum Yang-Baxter equation (QYBE)

$$R^{12}R^{13}R^{23} = R^{23}R^{13}R^{12}.$$

Moreover $\mathcal{O}(A) = \{P \in A \otimes_K A \mid (1 \otimes a)P = P(a \otimes 1)\}$ is a free rank one left (right) A-submodule of $A \otimes_K A$ with basis $\{Q\}$.

Let $Z(A)$ denote the center of the Frobenius K-algebra A; define $\Phi : A \to Z(A)$ by $\Phi(x) = \sum_{i=1}^n e^i x e_i$ for all $a \in A$; and set $u = \Phi(1)$.

Theorem ([39, Theorem 4.2]). *Consider the following conditions:*

(1) *u is invertible in A;*
(2) *A is a Frobenius $Z(A)$-algebra with Frobenius homomorphism Φ;*
(3) *$\Phi(x) = 1$ for some $x \in A$;*
(4) *A is a separable K-algebra.*

Then the implications $(1) \Rightarrow (2) \Leftrightarrow (3) \Leftrightarrow (4)$ hold.

In particular, if A satisfies one of conditions $(2), (3)$ or (4) then A is a separable K-algebra with separability idempotent $\sum e_i \otimes e^i x$, where x is the element of condition (3).

4.2. Hopf algebras as Frobenius algebras. Let H be a Hopf algebra over a commutative ring K with antipode S, comultiplication Δ, and counit ε. B. Pareigis showed that every Hopf algebra H that is finitely generated projective over a commutative ring K with trivial Picard group, is a Frobenius K-algebra with Frobenius homomorphism $\phi \in H^*$ such that ϕ satisfies

$$\sum_{(h)} h_1 \phi(h_2) = \phi(h) \cdot 1$$

for all $h \in H$, where $\sum_{(h)} h_1 \otimes h_2 = \Delta(h)$ is the comultiplication of h in Sweedler's notation. The latter condition on ϕ says that ϕ is an H-colinear map.

Recall that a left integral in H is an element $t \in H$ such that $ht = \varepsilon(h)t$ for all $h \in H$, and that H^* is also a Hopf algebra over K and denote its counit by π.

Theorem ([40, Theorem 3.2]). *Let H be a Hopf algebra over K which is Frobenius with a Frobenius homomorphism ϕ which is H-colinear. Then*

(1) $N = \sum_{i=1}^n \varepsilon(e_i)e^i$ *is a left integral in H and a left norm, i.e. $\phi(xN) = \varepsilon(x)$ for all $x \in H$.*

(2) ϕ *generates the submodule of left integrals in H^* and*

$$\mathrm{Tr}(S^2) = \varepsilon(N)\pi(\phi).$$

(3) *Given any left integral $l \in H$,*

$$R = (S^{-1} \otimes 1)\Delta(l)T \in \mathrm{End}(H \otimes_K H)$$

is a solution of QYBE.

The trace formula for the square of the antipode, which in the case that K is a field is due to *R. G. Larson* and *D. E. Radford* and was used in an essential way in their proof of a conjecture of Kaplansky (see [53, 54]).

Applying the last Theorem and the characterization of separable Frobenius algebras yields the following corollaries which generalize known results for Hopf algebras over fields:

Corollary ([40, Corollaries 3.4 and 3.5]). *Let H be a Hopf algebra over a commutative ring K such that H is finitely generated projective as K-module. Denote the antipode of H by S.*

(1) *H is a separable K-algebra if and only if H has a left integral l such that $\varepsilon(l)$ is an invertible element of K.*

(2) *H and H^* are separable K-algebras if and only if $\mathrm{Tr}(S^2)$ is an invertible element of K.*

A Hopf algebra H is called *involutory* if its antipode S is an involution, i.e., $S^2 = id$. As an extension of a theorem by Larson we have now:

Theorem. *The following conditions are equivalent for a Frobenius algebra H over an algebraically closed field K.*

   (a) *$u = \sum_{i=1}^{n} e_i e^i$ is invertible;*

   (b) *H is a separable K-algebra and the characteristic of K does not divide dimensions of simple H-modules.*

In particular, condition (a) is fulfilled if H is an involutory semisimple K-Hopf algebra.

If H is unimodular, i.e., the submodules of left integrals and of right integrals coincide, and finitely generated and projective as a K-module, then S acts as the identity on the submodule of integrals and $S^4 = id$ holds provided H^* is also unimodular.

5. Structure of matrix rings

Let n be a positive integer and R a ring, and let $M_n(R)$ denote the ring of $n \times n$ matrices over R.

5.1. CS matrix rings over local rings. In the papers [45, 46, 47, 48] with various coauthors Kostia made some contributions to the structure theory of CS modules and rings. A module M is called *CS* or *extending* provided every submodule of M is essential in a direct summand. M is said to be a *tight* module if every finitely generated submodule of the self-injective hull of M embeds in M. A ring R is called a *right CS-ring* if R is CS as a right R-module, and R is called *right tight* provided it is tight as a right module.

An open problem is to find necessary and sufficient conditions for direct sums of CS-modules to be CS. In [46] it is shown that if M is non-M-singular and CS, then M is M-tight and $\mathrm{End}(M_R)$ is right PP, and the converse also holds if M is furthermore a self-generator.

This result is applied to give necessary and sufficient conditions for R^n to be CS as a right R-module (equivalently, the $n \times n$ matrix ring $M_n(R)$ is a right CS-ring), where R is either a reduced ring or a ring with no infinite set of nonzero orthogonal idempotents. In particular, the open problem of characterizing a domain R such that R^2 is CS as a right R-module is solved; it is proved that such a domain is precisely a two-sided Ore domain and is two-sided 2-hereditary. Another result in this paper is:

Theorem. *For a von Neumann regular ring R, the following are equivalent for $n > 1$:*

(a) $M_n(R)$ is right weakly selfinjective;

(b) $M_n(R)$ is right $M_n(R)$-tight;

(c) $M_n(R)$ is a right CS-ring;

(d) R is right selfinjective.

In [45] a complete characterization of CS matrix rings $M_n(R)$, where $n > 1$, over local rings R is obtained:

Theorem. (1) $M_n(R)$ is right CS if and only if R is right uniform and for every right ideal I of R and for every R-homomorphism $f: I \to R$, there exists $a \in R$ such that either $f = L_a$ or $L_a f = \mathrm{id}_I$, where L_a is the left multiplication by a and id_I is the identity map on I.

(2) If, in addition, the Jacobson radical of R coincides with the right singular ideal $\{r \in R \mid rE = 0$ for some essential right ideal of $R\}$, then $M_n(R)$ is a right CS-ring if and only if R is selfinjective.

(3) If R is a commutative Noetherian local ring, then $M_n(R)$ is a right CS-ring if and only if the classical two-sided quotient ring, $Q(R)$, is a local QF-ring such that for all $q \in Q(R)$ either $q \in R$ or q is invertible in Q and $q^{-1} \in R$.

Applying the obtained results to group algebras, it is proved: If K is a field and G is a group (resp., nilpotent group) such that the group algebra KG is local (resp., semiperfect), then $M_n(KG)(n > 1)$ is a right CS-ring if and only if char$(K) = p$ and G is a finite p-group (resp., finite group). This result was subsequently generalized by the same authors in [47].

Theorem. *Let K be a field and G be a group. Suppose that one of the following conditions is satisfied:*

(i) *G is a locally finite group;*

(ii) *the group algebra KG is semilocal and G is either a solvable group or a linear group.*

Then the following conditions are equivalent:

(a) *$M_n(KG)$ for $n > 1$ is a right CS-ring;*

(b) *$M_2(KG)$ is a right CS-ring;*

(c) *KG is right self-injective;*

(d) *G is a finite group.*

5.2. Structure of right continuous right π-rings. A right module M is called π-*injective* or *quasi-continuous* if $f(M) \subseteq M$ for every idempotent $f \in \text{End}(E(M))$ where $E(M)$ is the injective hull of M. Quasi-continuous modules are in particular CS modules. A quasi-continuous module is called *continuous* if it is *direct injective*, i.e., if for every direct summand D of M every monomorphism $D \to M$ splits.

A ring R is called a right π-ring if every right ideal of R is π-injective. The structure of these rings was investigated in [48] leading to the following results.

For a positive integer n, let

(1) $D_1, D_2, \ldots, D_n$ be division rings,

(2) Δ be a right continuous right π-ring, all of whose idempotents are central, with essential ideal P such that Δ/P is a division ring and the right Δ-module Δ/P is not embeddable into Δ_Δ,

(3) V_i be a D_i-D_{i+1}-bimodule such that $\dim V_{iD_{i+1}} = 1$ for all $1 \le i < n$,

(4) V_n be a D_n-Δ-bimodule such that $V_n P = 0$ and $\dim V_{n\Delta/P} = 1$.

In this case, let $G_n(D_1, \ldots, D_n, \Delta, V_1, \ldots, V_n)$ denote the ring of $(n+1)$-by-$(n+1)$ matrices of the form

$$\begin{pmatrix} D_1 & V_1 & & & & \\ & D_2 & V_2 & & & \\ & & D_3 & V_3 & & \\ & & & \ddots & & \\ & & & & \ddots & \\ & & & & D_n & V_n \\ & & & & & \Delta \end{pmatrix}$$

with $V_i V_j = 0$ for all i, j.

The following result characterizes right continuous right π-rings.

Theorem. *A ring R is a right continuous right π-ring if and only if R is the direct sum of finitely many rings of the form*

$$G_n(D_1, \ldots, D_n, \Delta, V_1, \ldots, V_n),$$

finitely many indecomposable nonlocal right continuous right π-rings, and a right continuous right π-ring with all idempotents central.

5.3. Uniform bounds of primeness in matrix rings. A subset S of an associative ring R is a *uniform insulator* for R provided $aSb \ne 0$ for any nonzero $a, b \in R$. A ring R is called *uniformly strongly prime of bound $m(R)$* if R has uniform insulators and the smallest of these has cardinality $m(R)$. The systematic study of $m(R)$ was initiated by J. E. van den Berg who proved the following

Theorem. (1) *If F is an algebraically closed field, then $m(\text{M}_k(F)) = 2k - 1$.*

(2) *Let F be a field and assume there exists a nonassociative division F-algebra of dimension k, then $m(\mathbf{M}_k(F)) = k$.*

He asked if the converse of (2) holds. In [49, Theorem 1.2] a positive answer to this question is given showing how it is related to the existence of nonassociative division algebras over F.

Theorem. *Let F be a field and k a positive integer k. Then:*

(1) *$m(\mathbf{M}_k(F)) = 2k - 1$ for all k if and only if F is algebraically closed;*

(2) *$m(\mathbf{M}_k(F)) = k$ if and only if there exists a nonassociative division F-algebra of dimension k.*

5.4. Structure of rings with zero total. The *total* was introduced by F. Kasch, and Kostia was considering some questions arising from this notion. For two R-modules M and N, $\mathrm{Rad}(M, N)$ is defined as the set of all $g \in \mathrm{Hom}(M, N)$ such that $1 - fg$ is an automorphism of M for all $f \in \mathrm{Hom}(N, M)$. Let $\Delta(M, N)$ denote the set of all $g \in \mathrm{Hom}(M, N)$ such that the kernel of g is an essential submodule of M. Finally, let $\mathrm{Tot}(M, N)$ be the set of all $g \in \mathrm{Hom}(M, N)$ such that for all $f \in \mathrm{Hom}(N, M)$, $fg \neq (fg)^2$ unless $fg = 0$.

In [43], a joint paper of Kostia with F. Kasch, conditions on R and the modules are studied so that all three ideals are equal. In the special case $M = R$, the total R is the subset

$$\mathrm{Tot}(R) = \{a \in R : aR \text{ does not contain nonzero idempotents}\}.$$

In [38], Kostia considers rings with $\mathrm{Tot}(R) = 0$ and obtained the following

Theorem ([38, Theorem 5]). *Let $\mathrm{In}(a) = \min\{n \in \mathbb{N} \mid a^n = 0\}$ for a nilpotent element $a \in R$ and $\mathrm{In}(R) = \sup\{\mathrm{In}(a) \mid a \text{ nilpotent in } R\}$. For a ring R with $\mathrm{In}(R) = n < \infty$, the following are equivalent:*

(a) *$\mathrm{Tot}(R) = 0$;*

(b) *R contains an essential ideal I which is a direct sum of ideals $I_k = \mathbf{M}_{n_k}(D_k)$, $k = 1, 2, \ldots, t$, where*

    (1) *$n_1 < n_2 < \cdots < n_t = n$.*

    (2) *Each $\mathbf{M}_{n_k}(D_k)$ is a matrix ring over a reduced ring D_k.*

    (3) *If $L \neq 0$ is a right ideal of D_k, then the set of all central idempotents of D_k belonging to L generates an essential ideal in L.*

This result yields interesting corollaries, e.g.

Corollary ([38, Corollary 6]). *The following statements are equivalent for a ring R with $\mathrm{In}(R) = n < \infty$:*

(a) *R is a prime ring and $\mathrm{Tot}(R) = 0$.*

(b) *$R \cong M_n(D)$, where D is a division ring.*

Corollary ([38, Corollary 8]). *Let R be a ring with $\mathrm{Tot}(R) = 0$ and $\mathrm{In}(R) = n < \infty$. Then the following statements hold:*

(1) *The maximal right ring of quotients Q of R equals the maximal left ring of quotients of R.*

(2) *Q is isomorphic to a finite direct sum of matrix rings over abelian regular left and right self-injective rings.*

References

[1] Beidar, K. I. *Rings with generalized identities. I.* (Russian) Vestn. Mosk. Univ., Ser. I **1977** (1977), No.2, 19–26.

[2] Beidar, K. I. *Rings with generalized identities. I.* Mosc. Univ. Math. Bull. **32** (1977), No. 2, 15–20.

[3] Beidar, K. I. *Semiprime rings with generalized identity.* (Russian) Usp. Mat. Nauk **32** (1977), No. 4(196), 249–250.

[4] Beidar, K. I. *Rings with generalized identities. II.* Mosc. Univ. Math. Bull. **32** (1977), No. 3, 27–33.

[5] Beidar, K. I. *The ring of invariants under the action of a finite group of automorphisms of a ring.* (Russian) Usp. Mat. Nauk **32** (1977), no. 1(193), 159–160.

[6] Beidar, K. I. *Associative rings and finite groups of automorphisms.* (Russian) Tr. Semin. Im. I.G. Petrovskogo **4** (1978), 33–44.

[7] Beidar, K. I. *Rings with generalized identities III.* Mosc. Univ. Math. Bull. **33** (1978), No. 4, 53–58.

[8] Beidar, K. I. *Rings of quotients of semiprime rings.* (Russian) Vestn. Mosk. Univ., Ser. I **1978** (1978), No. 5, 36–43.

[9] Beidar, K. I. *Quotient rings of semiprime rings.* Mosc. Univ. Math. Bull. **33** (1978), No. 5, 29–34.

[10] Beidar, K. I. *Classical quotient rings of PI-algebras.* Russ. Math. Surv. **33** (1978), No. 6, 223–224.

[11] Beidar, K. I.; Mikhalev, A. V.; Salavova, K. *Generalised identities and semiprime rings with involution.* (Russian) Usp. Mat. Nauk **35** (1980), No. 1(211), 222.

[12] Beidar, K. I.; Mikhalev, A. V.; Salavova, K. *Generalised identities and semiprime rings with involution.* Russ. Math. Surv. **35** (1980), No. 1, 209–210.

[13] Beidar, K. I.; Mikhalev, A. V.; Salavova, K. *Generalised identities and semiprime rings with involution.* Math. Z. **178** (1981), 37–62.

[14] Beidar, K. I. *On modules over commutative semi-primary rings.* (Russian) Mat. Zametki **29** (1981), 15–18.

[15] Beidar, K. I. *Modules over commutative semiprime rings*. Math. Notes **29**, 9–10 (1981).

[16] Beidar, K. I.; Mikhalev, A. V. *The functor of orthogonal completion*. (Russian) Abelevy Gruppy Moduli **4**, 3–19 (1986).

[17] Beidar, K. I.; Mikhalev, A. V. *Orthogonal completeness and minimal prime ideals*. (Russian. English summary) Tr. Semin. Im. I.G. Petrovskogo **10**, 227–234 (1984).

[18] Beidar, K. I.; Mikhalev, A. V. *Orthogonal completeness and minimal prime ideals*. J. Sov. Math. **35**, 2876–2882 (1986).

[19] Beidar, K. I.; Mikhalev, A. V. *Orthogonal completeness and algebraic systems*. (Russian, English) Russ. Math. Surv. **40** (1986), No.6, 51–95; translation from Usp. Mat. Nauk **40** (1985), No.6 (246), 79–115.

[20] Beidar, K. I.; Mikhalev, A. V.; Slin'ko, A. M. *A criterion for non-degenerate alternative and Jordan algebras to be prime*. (Russian) Tr. Mosk. Mat. O.-va **50** (1987), 130–137.

[21] Beidar, K. I.; Mikhalev, A. V. *Structure of nondegenerate alternative algebras*. (Russian. English summary) Tr. Semin. Im. I. G. Petrovskogo **12** (1987), 59–74.

[22] Beidar, K. I.; Mikhalev, A. V.; Slin'ko, A. M. A criterion for primeness of nondegenerate alternative and Jordan algebras. (Russian, English) Trans. Mosc. Math. Soc. **1988** (1988), 129–137; translation from Tr. Mosk. Mat. O.-va **50** (1987), 130–137.

[23] Beidar, K. I.; Mikhalev, A. V. *Semiprime rings with bounded indexes of nilpotent elements*. (Russian. English summary) Tr. Semin. Im. I. G. Petrovskogo **13** (1988), 237–249.

[24] Beidar, K. I.; Salavova, K. *On a class of rings with essential right socle*. (Russian) Acta Math. Hung. **53** (1989), No. 1/2, 55–59.

[25] Beidar, K. I.; Mikhalev, A. V. *Structure of nondegenerate alternative algebras*. (Russian, English) J. Sov. Math. **47** (1989), No. 3, 2525–2536; translation from Tr. Semin. Im. I. G. Petrovskogo 12 (1987), 59–74.

[26] Beidar, K. I.; Mikhalev, A. V. *Semiprime rings with bounded indices of nilpotent elements*. (Russian, English) J. Sov. Math. **50** (1990), No. 2, 1518-1526; translation from Tr. Semin. Im. I. G. Petrovskogo 13 (1988), 237–249.

[27] Beidar, K. I.; Trokanová-Salavová, K. *On nil rings satisfying minimum condition on principal right ideals*. Acta Math. Hung. **55** (1990), No. 3/4, 197–200.

[28] Beidar, K. I. *The Andrunakievich lemma and Jordan algebras*. (Russian, English) Russ. Math. Surv. **45** (1990), No. 4, 159; translation from Usp. Mat. Nauk **45** (1990), No. 4(274), 137–138.

[29] Beidar, K. I.; Mikhalev, A. V. *Generalised polynomial identities and rings which are sums of two subrings*. (Russian, English) Algebra Logic 34, No. 1, 1–5 (1995); translation from Algebra Logika **34** (1995), No. 1, 3–11.

[30] Beidar, K. I. *Classical localizations of alternative algebras*. (Russian. English summary) Tr. Semin. Im. I. G. Petrovsk. **16** (1992), 227–235.

[31] Beidar, K. I. *Classical localizations of alternative algebras*. (Russian, English) J. Math. Sci., New York **69**, No. 3, 1098–1104 (1994); translation from Tr. Semin. Im. I. G. Petrovsk. **16** (1992), 227–235.

[32] Beidar, K. I. *Classical rings of quotients of semiprime PI-rings*. (Russian, English) Algebra Logic **32**, No. 1, 1–8 (1993); translation from Algebra Logika **32** (1993), No. 1, 3–16.

[33] Beidar, K. I.; Markov, V. T. *A semiprime PI-ring having a faithful module with Krull dimension is a Goldie ring.* (Russian, English) Russ. Math. Surv. **48** (1993), No. 6, 158; translation from Usp. Mat. Nauk **48** (1993), No. 6 (294), 141–142.

[34] Beidar, K. I.; Wisbauer, R. *Strongly and properly semiprime modules and rings.* Jain, S. K. (ed.) et al., Ring theory. Proceedings of the biennial Ohio State-Denison mathematics conference, May 14–16, 1992 dedicated to the memory of H. J. Zassenhaus. Singapore: World Scientific (1993). pp. 58–94.

[35] Beidar, K. I.; Wisbauer, R. *Strongly semiprime modules and rings.* (Russian, English) Russ. Math. Surv. **48** (1993), No. 1, 163–164; translation from Usp. Mat. Nauk. **48** (1993), No. 1(289), 161-162.

[36] Beidar, K. I.; Wisbauer, R. Properly semiprime self-pp-modules. Commun. Algebra **23** (1995), No. 3, 841–861.

[37] Beidar, K. I.; Mikhalev, A. V. *The method of orthogonal completeness in the structure theory of rings.* J. Math. Sci., New York **73** (1995), No. 1, 1–46.

[38] Beidar, K. I. *On rings with zero total.* Beitr. Algebra Geom. **38** (1997), No. 2, 233–239.

[39] Beidar, K. I.; Fong, Y.; Stolin, A. *On Frobenius algebras and quantum Yang-Baxter equation.* Trans. Am. Math. Soc. **349** (1997), No. 9, 3823–3836.

[40] Beidar, K. I.; Fong, Y.; Stolin, A. *On antipodes and integrals in Hopf algebras over rings and the quantum Yang-Baxter equation.* J. Algebra **194** (1997), No. 1, 36–52.

[41] Beidar, K. I.; Mikhalev, A. V.; Puninskij, G. E. *Logical aspects of the theory of rings and modules.* (Russian. English summary) Fundam. Prikl. Mat. **1** (1995), No. 1, 1–62.

[42] Beidar, K. I.; Mikhalev, A. V. *Antiisomorphisms of endomorphism rings of modules close to free ones which are induced by Morita antiequivalences.* (Russian, English) J. Math. Sci., New York **85** (1997), No. 6, 2450-2453; translation from Tr. Semin. Im. I. G. Petrovskogo **19** (1996), 338–344.

[43] Beidar, K. I.; Kasch, F. *Good conditions for the total.* Birkenmeier, Gary F. (ed.) et al., Proc. 3rd Korea-China-Japan symposium, Korea, 1999. Boston, MA: Birkhäuser (2001). Trends in Mathematics. pp. 43–65.

[44] Beidar, K. I.; Torrecillas, B. *On actions of Hopf algebras with cocommutative coradical.* J. Pure Appl. Algebra **161** (2001), No. 1-2, 13–30.

[45] Beidar, K. I.; Jain, S. K.; Kanwar, P.; Srivastava, J. B. *CS matrix rings over local rings.* J. Algebra **264** (2003), No. 1, 251–261.

[46] Beidar, K. I.; Jain, S. K.; Kanwar, P. *Nonsingular CS-rings coincide with tight PP rings.* J. Algebra **282** (2004), No. 2, 626–637.

[47] Beidar, K. I.; Jain, S. K.; Kanwar, P.; Srivastava, J. B. *Semilocal CS matrix rings of order > 1 over group algebras of solvable groups are selfinjective.* J. Algebra **275** (2004), No. 2, 856–858.

[48] Beidar, K. I.; Jain, S. K. *The structure of right continuous right π-rings.* Commun. Algebra **32** (2004), No. 1, 315–332.

[49] Beidar, K. I.; Wisbauer, R. *On uniform bounds of primeness in matrix rings.* J. Aust. Math. Soc. **76** (2004), No. 2, 167–174.

[50] Handelman, D.; Lawrence, J. *Strongly prime rings.* Trans. Am. Math. Soc. **211** (1975), 209–223.

[51] Handelman, D. *Strongly semiprime rings.* Pac. J. Math. **60** (1975), 115–122.

[52] Erickson, T. S.; Martindale, W. S. III; Osborn, J. M. *Prime nonassociative algebras.* Pac. J. Math. **60** (1975), 49–63.

[53] Larson, R. G.; Radford, D. E. *Finite dimensional cosemisimple Hopf algebras in characteristic 0 are semisimple.* J. Algebra **117** (2) (1988), 267–289.

[54] Larson, R. G.; Radford, D. E. *Semisimple cosemisimple Hopf algebras.* Am. J. Math. **110** (1) (1988), 187–195.

[55] Lomp, C. *A central closure construction for certain algebra extensions. Applications to Hopf actions.* J. Pure Appl. Algebra **198** (2005), No. 1-3, 297–316.

[56] Martindale, W. S. III. *Prime rings satisfying a generalized polynomial identity.* J. Algebra **12** (1969), 576–584.

[57] Martindale, W. S. III. *On semiprime P.I. rings.* Proc. Am. Math. Soc. **40** (1973), 365–369.

[58] Wisbauer, R. *Localization of modules and the central closure of rings.* Commun. Algebra **9** (1981), No. 14, 1455–1493.

Christian Lomp, Centro de Matemática da Universidade do Porto, Rua Campo Alegre, 687, 4169-007 Porto, Portugal
E-mail address: clomp@fc.up.pt

Robert Wisbauer, Mathematisches Institut der Heinrich-Heine Universität Düsseldorf, Universitätsstraße 1, 40225 Düsseldorf, Germany
E-mail address: wisbauer@math.uni-duesseldorf.de

Lie maps in prime rings: a personal perspective

Wallace S. Martindale, 3rd

Dedicated to the memory of Konstantin I. Beidar.

0. Introductory Remarks

As an attendee at the conference held at National Cheng-Kung University (Tainan) during March 2005 in honor of and in memory of Konstantin Beidar, I was asked to submit to the Conference Proceedings a paper on Lie maps, a topic to which Beidar made very substantial contributions. I am taking the liberty of writing this paper in the first person singular (using "I") rather than in the first person plural (the royal "we"). Somehow this more personal touch seems appropriate under the circumstances; Beidar was a close personal friend as well as a wonderful mathematical colleague.

Although I knew of Beidar before 1989 (the familiar phrase "by an example due to Beidar" had several times come to my attention), it was at an international algebra conference at Novosibirsk in that year that I was introduced to him by his mentor Alexander Mikhalev. At that time his spoken English was very minimal; over the years since then, although there remained a strong accent, it improved immensely. It was there that Mikhalev and Beidar asked me if I would be interested in writing a book with them on rings satisfying various generalized identities (see [14]) and I recall eagerly replying in the affirmative. As a result there then followed four visits to Moscow where I generally stayed at the University of Moscow, with Beidar looking out for my well-being way beyond the call of duty. Several years later, when he and his family had moved to Tainan, I was again the recipient of enormous hospitality from Beidar and his family, living in their apartment during each of several visits. My heart goes out to his widow Lida and daughters Katya and Tanya, now living in New Zealand where Beidar had moved his family.

In a natural way this paper is divided into two parts. Up to the path-breaking theorem of Brešar on Lie isomorphisms of prime rings [16, Theorem 3], which appeared in 1993, all the results (which I know about) on Lie maps made use of the presence of non-trivial idempotents. Many of these are due to myself and to some of my students. I will discuss these in section 1. Brešar's theorem set the stage for a host of theorems which ultimately answered in full generality all of the Lie map problems posed by Herstein in his 1961 AMS Hour talk [19]. It is here that Beidar (along with Brešar and Chebotar) was the driving force, putting to use his and Chebotar's theory of functional identities and d-freeness. I will mention some aspects of this in section 2.

Before proceeding further I will just make a few general comments. If A and R are associative rings, then any map $\sigma : A \rightarrow R$, where σ is either a homomorphism or the negative of an anti-homomorphism, clearly preserves commutators (i.e., preserves the Lie product $[x, y]$). But also any additive map $\tau : A \rightarrow Z$ which vanishes on commutators, where Z is the center of R, is a Lie homomorphism. Conversely, now suppose $\alpha : L \rightarrow R$ is a Lie homomorphism, where L is a Lie subring of A (e.g. L is A itself or, if A has an involution, L is the set of skew elements of A). Then, if A and/or R is a properly conditioned ring, one would hope to show that α is induced by some combination of the maps σ and τ given above. For the rest of this paper I will always assume char. $\neq 2$. I will often just use the term *ring* (which is an algebra over the integers) rather than the more general term *algebra over a commutative ring with* 1.

1. Lie maps under idempotent assumptions

My introduction to the problem of characterizing Lie isomorphisms (in associative rings) began when I became a student of Herstein at the University of Pennsylvania in 1953. Herstein, who was in the process of developing his theory of Lie and Jordan structures in simple rings (with and without involutions), suggested the following thesis problem for me:

Conjecture 1. *If α is a Lie isomorphism of a simple ring A onto a simple ring R, then $\alpha = \sigma + \tau$, where $\sigma : A \rightarrow R$ is either an isomorphism or the negative of an antiisomorphism and τ is an additive mapping of A into the center Z of R vanishing on commutators.*

I distinctly recall making absolutely zero progress and simply gave up; I settled instead for writing my dissertation on two of the myriad of generalizations of Jacobson's famous "$x^n = x$ implies commutativity" theorem (which had given rise to Herstein's commutativity theorems).

During my first position as instructor at the college of the University of Chicago in the late 1950's, finally realizing that it might be a good idea if I tried to prove something, I thought back to the Lie isomorphism conjecture. I vividly remember walking into Kaplansky's office and asking if he had any suggestions about how to tackle this problem. His reply was: why not assume that there are at hand an infinite number of orthogonal idempotents? This advice proved a godsend to me (it subsequently turned out that three orthogonal idempotents were enough). Fortunately about this same time I started to study Hua's 1951 result [21] in which he proved Conjecture 1 in the special case that $A = R = M_n(D)$, where $n \geq 3$ and D is an arbitrary division ring. Hua's paper may well have been the first paper in which Lie isomorphisms were treated in this degree of generality. There were some very useful equations in Hua's paper showing how the Lie automorphism acted on idempotents and on pairs of orthogonal idempotents. So it soon occurred to me that a natural way of generalizing Hua's result was to assume A and R were primitive rings in which A contained three orthogonal idempotents whose sum was 1. Then one could write A in its Pierce decomposition with respect to the three idempotents, then just "pretend" that A behaved in many ways like 3×3 matrices, and hope for the best! Fortunately my efforts were successful here [24], although one "wrinkle" appeared in the conclusion of the theorem which proved to be an omen for the future. Namely, the image ring R, being primitive, is by definition a "dense" subring of the full ring of linear transformations T of a vector space over a division ring, with the center Z of R contained in the center C of T. For the purposes of this paper I will refer to T as the *primitive closure* of the primitive ring R. I was able to prove

Theorem 1. *Let A, R be primitive rings, let T be the primitive closure of R, let C be the center of T, and suppose A has three orthogonal idempotents whose sum is 1. If $\alpha : A \to R$ is a Lie isomorphism then $\alpha = \sigma + \tau$, where $\sigma : A \to T$ is either a monomorphism or the negative of an anti-monomorphism and $\tau : A \to C$ is an additive mapping vanishing on commutators.*

An easy example [24, p. 916] illustrated the need for enlarging the image ring. Since simple rings containing an idempotent must be primitive, Theorem 1 answered Conjecture 1 under the idempotent assumptions. My first student, R.A. Howland, considered a Lie isomorphism $\alpha : [A, A] \to [R, R]$ where A and R were simple rings. Under the same idempotent assumptions as in Theorem 1 he was able to show that α could be extended to either an isomorphism or the negative of an anti-isomorphism of A onto R. It should be noted in this situation that one does not have the luxury of $[A, A]$ being closed under any "associative" product (such as xy or even x^2 or x^3); however the Pierce decomposition could still be used effectively.

Results on isomorphisms (or more generally homomorphisms) naturally suggest related results on derivations in view of the following well-known observation. Let $\delta : A \to A$ be a derivation and let R denote the ring of all upper triangular 2×2 matrices over A. Then the map $\alpha : A \to R$ given by

$$x \mapsto \begin{pmatrix} x & \delta(x) \\ 0 & x \end{pmatrix} \tag{1}$$

is a homomorphism. This observation holds for any arbitrary non-associative ring, in particular, for Lie derivations of an associative ring A regarded as a Lie ring under $[x, y]$. At first glance it would seem that derivation results would tumble out from Lie homomorphism results "free of charge", but the downside is that the image of α lies in a ring which may no longer be suitable; in section 2 it will be indicated (Theorem 7) how this barrier can be circumvented in some rather general situations). One does have the feeling, however, that derivations are in a sense easier to derive results for than are homomorphisms, perhaps partly because the domain and range of the derivation interact. This was born out when I proved the companion result of Theorem 1 for Lie derivations [25].

Theorem 2. *Let R be a primitive ring containing an idempotent $e \neq 0, \neq 1$ and let T be the primitive completion of R with center C. If $\gamma : R \to R$ is a Lie derivation, then $\gamma = \delta + \tau$, where $\delta : R \to T$ is a derivation and $\tau : R \to C$ is an additive mapping vanishing on commutators.*

Let's backtrack a bit at this point. The classification of finite dimensional simple Lie algebras over a field of characteristic 0 had already been well worked out (see, e.g., [22, Chapter 10]). The four "great classes" of simple Lie algebras L arose from matrix algebras $R = M_n(F)$, with F a field of characteristic 0: $L = [R, R]$ (type A) and, if R has an involution * of the first kind (i.e. F elementwise fixed under *), $L = [K, K]$ is the skew elements of R under * (type B, C, or D). If * is of the second kind (i.e. F contains a nonzero skew element) then L can be shown to be of type A. Thus the study of Lie algebras serves as some motivation for studying associative rings with involution in general. The following question naturally arises: how does one tell if two simple Lie algebras are isomorphic? In the classical situation being discussed here it was shown (again see [22, Chapter 10]), that any Lie isomorphism of two simple Lie algebras can be uniquely extended (modulo some low dimensional counterexamples) to an isomorphism of the corresponding central simple associative algebras from which the Lie algebras were obtained. Thus we have some motivation for studying Lie isomorphisms in general.

Herstein was convinced that the simplicity of the simple Lie algebras just described should follow solely from the fact that $R = M_n(F)$ is a *simple* ring. This led him in the 1950's to develop his Lie theory of arbitrary simple rings (without and with involutions), and his efforts in this direction were honored with an invitation to address the American Mathematical Society in an Hour Talk in 1961 [19]. In that lecture he summarized his results on the Lie structure of simple rings: (1) if R is simple with center Z then $[R, R]/[R, R] \cap Z$ is a simple Lie ring and (2) if R is a simple ring with involution * and K denotes the skew elements under *, then $[K, K]/[K, K] \cap Z$ is a simple Lie ring. An example due to P.-H. Lee shows that even if * is of the first kind one cannot assume that $K = [K, K]$ (in contrast to the classical situation above). Also in general $[R, R]$ and $[K, K]$ may intersect the center Z. For instance, if $R = M_n(F)$, where characteristic F divides n, then $[R, R] \cap Z \neq 0$. Hence the simple Lie rings may turn out to be factor rings.

In this same talk Herstein proposed a series of open questions aimed at characterizing Lie isomorphisms in simple rings: if two of the simple Lie rings indicated above are isomorphic as Lie rings, can these Lie isomorphisms be essentially lifted to (anti)-isomorphisms of the simple associative rings from which the simple Lie rings arose? Additionally included were questions about Lie isomorphisms where the Lie ring was "not quite" simple, e.g. see Conjecture 1 above. Analogous problems for Lie derivations were also raised.

About this same time prime rings (the ultimate generalization of simple and primitive rings) were coming into vogue, no doubt due in part to Goldie's famous result characterizing prime Noetherian rings. A student of mine, T. E. Erickson, generalized Herstein's Lie theory of simple rings to prime rings [18]. It seemed appropriate now to phrase Herstein's Lie isomorphism questions in the context of prime rings, with obvious adjustments such as the object now being to show that the associative subring generated by the respective prime Lie rings were (anti)-isomorphic. For future reference we let $\langle S \rangle$ denote the subring generated by a subset S of a ring.

The natural first step to take now was to consider a Lie isomorphism $\alpha : A \rightarrow R$ where A and R are prime rings. Following the lead of Theorem 1 above I assumed that A contained orthogonal idempotents e_1, e_2 whose sum was 1, thereby being able to work in the framework of the Pierce decomposition. However, an obstacle immediately presented itself, namely, I knew from the example illustrating Theorem 1 that the image of τ could not be assumed to lie in the center Z of R. But whereas the case in which R was primitive allowed one to use the center of the primitive closure of R, for prime rings there did not at first seem to be an obvious candidate for an extension field of Z which would accommodate the image of τ. However, Lambek's book [23] had just appeared, and his account of Utumi's description of the so-called maximal right ring of quotients Q of R came to the rescue, namely, the center C of Q turned out to be just the extension field of Z I was seeking: $\alpha = \sigma + \tau$, with $\sigma(a) \subseteq RC$ and $\tau(a) \subseteq C$ [26, Theorem 11]. The key lemma in the proof was

Lemma 1. *Let R be prime and let $a, b \in R$ such that $axb = bxa$ for all $x \in R$. Then a and b are $C - dependent$.*

My inspiration for the proof of this lemma was a combination of Utumi's description of the maximal ring of right quotients in concert with a lemma from Amitsur's fundamental 1965 paper [1, Lemma 6a], characterizing primitive rings satisfying a generalized polynomial identity (GPI). So what was now occurring was the pleasant phenomenon of the study of one topic (i.e. Lie isomorphisms) leading me inexorably into the study of another topic (i.e. GPI's). In view of Kaplansky's theorem on primitive PI rings, Posner's theorem on prime PI rings, and Amitsur's theorem on primitive GPI rings, what else could a person do than try to characterize prime GPI rings (which I did in a 1969 paper [27] in which Lemma 1 was again the key tool). In this paper I gave the name *extended centroid* to the center of Q and the name *central closure* to RC. Shortly thereafter [28] I showed that if R was a prime ring with involution it followed that if the skew elements K of R under * were GPI then R itself must be GPI. These GPI results would in turn be helpful in the further study of Lie isomorphisms.

In passing I remark that there had evolved in functional analysis some interest in Lie isomorphisms and derivations, no doubt due to the fact that there is an abundance of idempotents in many operator algebras. As instances I will just cite the work of Ayupov [2] and Miers [30, 31].

I can't remember who told me that for a theorem to be a decent one there should be some counterexamples! Of course such a remark can't be taken too seriously, but the point is that if there are some cases which one must avoid then the proof can't be completely a formal one. My 1976 result [29], in which I settled one of Herstein's conjectures for prime rings with involution of the first kind under the assumption of three orthogonal symmetric idempotents, I believe qualified in this respect.

Theorem 3. *Let A and R be prime rings with involutions of the first kind, with respective extended centroids D and C, and respective skew elements L and K. Suppose that $(AD:D) \neq 1, 4, 9, 16, 25, 64$ and that AD contains orthogonal symmetric idempotents e_1, e_2 such that $e_1 + e_2 \neq 1$ (plus another "technical" condition on the e_i's which I won't mention here). Then any Lie isomorphism $\alpha : [L, L] \to [K, K]$ can be extended uniquely to an associative isomorphism of $\langle [L, L] \rangle$ onto $\langle [K, K] \rangle$.*

There are indeed counterexamples illustrating each of the dimensions given above.

It was natural to next investigate the situation in which A and R were prime rings with involution of the second kind; thus one is given a Lie isomorphism

$$\alpha : [L, L]/[L, L] \cap D \longrightarrow [K, K]/[K, K] \cap C,$$

the aim being to find a ring monomorphism $\sigma : \langle [L, L] \rangle \to RC$ such that $\overline{x^\sigma} = x^\alpha$ for all $x \in [L, L]$. The appropriate result here was obtained by my student M. P. Rosen in 1984 [32, Theorem 2.1] under the same idempotent assumptions as above and under the additional assumption that neither A nor R was GPI. Much later, in 1998, in a joint paper with my student P. S. Blau, Rosen's theorem was obtained under the assumption that at least one of A and R was GPI (the idempotent conditions follow from the GPI property).

In the early 1990's (I don't remember the exact date) I received a telephone call from Charles Lanski, who said he had just refereed a paper [16] written by some guy from Slovenia who had just proved

Theorem 4. *If A and R are prime rings, with neither A nor R satisfying the standard identity S_4, and $\alpha : A \to R$ is a Lie isomorphism, then $\alpha = \sigma + \tau$ where $\sigma : A \to RC + C$ is either a ring monomorphism or the negative of a ring anti-monomorphism and $\tau : A \to C$ is an additive map vanishing on commutators.*

The analogous theorem for Lie derivations of a prime ring R was also worked out in the same paper [16].

Wow! Goodbye idempotents, I thought. Well, it was a good run while it lasted. And this person was a functional analyst, not even a ring theorist; he is now, I knew! His name was Matej Brešar, and he was responsible for ushering in the idempotent-free era of Lie isomorphism results, which forms the subject of section two of this paper.

2. The complete solution of Herstein's problems

Shortly after the advent of Brešar's result (Theorem 4 above) my colleague C.
R. Miers (mentioned earlier) got Brešar and me together at the University of Victoria, where the three of us collaborated on several papers. At this same time I was
already periodically visiting Moscow to work with Beidar (and Mikhalev) on our
book on generalized identities [14]. It didn't take long to appreciate Beidar's fertile imagination and broad knowledge of the ring theory literature. What I want to
focus on here is a particular occasion during one of my Moscow visits shortly after
the appearance of Brešar's result. Of course I had wanted to prove the analogue
of Brešar's theorem for the more difficult case of a Lie isomorphism $\alpha : L \to K$,
where L, K were the respective skew elements of prime rings A, R with involutions of the first kind. However, I was reticent about diverting our attention from
the book by plunging into the murky waters of a difficult side problem. But one
morning Beidar suddenly said in words to this effect, "Let's forget about the book
for a while now; I think I have a plan on how to prove the Lie isomorphism theorem". It was, of course, pretty obvious to both of us that the only way to get started
was to try to find out how α acted on cubes of elements of L; it was Beidar who
worked out the details [13, Lemma 8], of showing that preserving cubes was actually equivalent to showing that α could be lifted to an isomorphism of $\langle L \rangle$ onto
$\langle K \rangle$. The brunt of the proof, of course, was to show that α preserved cubes. Let
$\{x, y, z\} = xyz + xzy + yxz + yzx + zxy + zyx$. From linearizing $[(x^3)^\alpha, x^\alpha] = 0$
and setting $B(x, y, z) = \{x, y, z\}^\alpha$ one easily sees that

$$[B(x, y, z), t^\alpha] + [B(y, z, t), x^\alpha] + [B(z, t, x), y^\alpha] + [B(t, x, y), z^\alpha] = 0 \quad (2)$$

for all $x, y, z, t \in K$. [Remark: Certainly Beidar was to see (2) as a prime motivating example for what would later become the basic type of "functional identity"
in the "d-freeness" theory soon to be developed by him and the young Russian
Mikhail Chebotar (whom I first met at a conference in Hungary).] We showed that
(2) alone was enough to prove that

$$B(x, y, z) \quad = \quad \lambda\{x^\alpha, y^\alpha, z^\alpha\} + \mu(x, y)z^\alpha + \mu(x, z)y^\alpha + \mu(y, z)x^\alpha \quad (3)$$

for a fixed $\lambda \in C$ and a symmetric bilinear map $\mu : K^2 \to C$ (note that 0-degree
and 2nd-degree terms don't appear because they are symmetric elements of R).
[Remark: expressions such as (3) were to play an important role in the Beidar-Chebotar theory of functional identies and were to be called *quasi polynomials*.]
The additional properties of α being a Lie isomorphism were then employed to
show that $\lambda = 1$ and $\mu = 0$. Since GPI rings lead to plenty of idempotents, in

view of Theorem 3 we could assume that A (and hence R) were non-GPI rings. [Remark: thank goodness that idempotents still had their usefulness, I thought to myself, in handling "low-dimensional" cases.] As we were making many calculations involving various elements x^α, y^α, etc. I remember very clearly Beidar saying that in order to have "control" over what was happening it was important that the subalgebra P over C generated by certain of these elements was partially "free" in the sense (very roughly speaking) that the subspace of P of elements of "degree" less than some fixed m was free. [I believe this "freeness" that we required in our paper in some way evolved into the Beidar-Chebotar notion of "d-freeness" later on]. It was Beidar who came up with the following key lemma [13, Lemma 1] on non-GPI rings which was used over and over again in making our proofs work. Let R be a closed prime ring with involution (i.e. R is its own central closure), and let $R_C\langle X \rangle$ denote the free product of R and the free algebra $C\langle X \rangle$ over C.

Lemma 2. *Suppose R is not GPI. Let $T_i = \{f_{ij}(X) \in R_C\langle X \rangle \mid j = 1, 2, \ldots, n_i\}$, $i = 1, 2, \ldots, m$, be m given subsets of $R_C\langle X \rangle$, each of which is C-independent. Then there exists $x \in K$ such that, for each $i = 1, 2, \ldots, m$,*

$$T_i(x) = \{f_{ij}(x) \mid j = 1, 2, \ldots, n_i\}$$

is a C-independent subset of R.

Here's a statement of our main result:

Theorem 5. *Let A and R be prime rings with involutions of the first kind of char. $\neq 2, 3$, with L and K the respective skew elements of A and R, and D and C the respective extended centroids of A and R. If $(AD : D) \neq 1, 4, 9, 16, 25, 64$ then any Lie isomorphism $\alpha : L \to K$ can be extended uniquely to an associative isomorphism of $\langle L \rangle$ onto $\langle K \rangle$.*

A couple of isolated footnotes: Chebotar [17] was able to remove the assumption of characteristic $\neq 3$ in Theorem 5, my student G. Swain [33] proved the analogue of Theorem 5 for derivations, and my student P. Blau [15] settled the case in Theorem 4 where either A or R satisfies S_4.

With Brešar having pioneered the use of functional identities and with the preceding theorem as well as a host of other results in mind, Beidar and Chebotar, in two fundamental papers [7, 8], took a major unifying and generalizing step forward by introducing the notion of *d-freeness* and then developing a cohesive theory of functional identities (sometimes to be abbreviated as FI's). By its very nature this theory is fraught with notational complications, so I shall content myself with illustrating in a very informal way the notion of d-freeness by means of a rather specific example (for a complete explanation see, e.g., the article by Brešar in these Conference Proceedings). A typical framework in which d-freeness operates (especially one conducive to Lie isomorphisms) would be the situation in which we have a map $\alpha : S \rightarrow Q$, where S is a set and Q is a unital ring with center C. For example, in Theorem 5 above Q is the maximal right ring of quotients of the prime ring R with involution, S is the skew elements of A, and α is a Lie isomorphism of S onto the skew elements of R. For our example let $E, F, G, H : S^2 \rightarrow Q$ be (unknown) functions, let $\alpha : S \rightarrow Q$ be a given function, and consider the following example of a *basic* "tri-additive" functional identity:

$$E(x, y)z^\alpha + F(x, z)y^\alpha \;\; = \;\; y^\alpha G(x, z) + z^\alpha H(x, y) \tag{4}$$

for all $x, y, z \in S$. What are the solutions for E, F, G, H? Some "obvious" ones are staring us in the face: let $p, q : S \rightarrow Q$ and $\lambda : S^2 \rightarrow C$ be *any* functions, and set $E(x, y) = y^\alpha p(x)$, $F(x, z) = z^\alpha q(x) + \lambda(x, z)$, $G(x, z) = p(x)z^\alpha + \lambda(x, z)$, $H(x, y) = q(x)y^\alpha$. The reader could perhaps extrapolate from this example what in general would be the form of a basic m-additive FI and what the "obvious" solutions would be. Very loosely and even inaccurately put, the pair (S, α) is said to be *d-free* if all basic d-additive FI's have *only* the "obvious" solutions. In case $S \subseteq Q$ and $\alpha = 1$ one simply says that S is d-free. Thus in the above example, if one happens to know that (S, α) is 3-free then any solution of (4) is of the form indicated above. In this example, of course, E, F, G, H cannot be further simplified. Fortunately, however, there are situations (for example, such as equation (2) in which all the unknown functions are the same function B) in which (under a suitable d-freeness assumption) much more can be said, the ultimate hope being that the unknown functions turn out to be so-called quasi-polynomials (as illustrated in (3) above). In true Russian fashion of generalizing to the nth degree, all these matters were addressed in great detail and thoroughness by Beidar and Chebotar in [7] and [8]; I cannot overemphasize the importance of these two fundamental papers. It would be an understatement to say that these two papers are not "bedtime" reading; nevertheless what might at first appear to be excessively general and complicated has in fact often turned out to be what is needed for various applications (e.g., settling Herstein's Lie map conjectures).

Of course, one might worry that d-freeness is just an empty notion, just an artificial invention designed to solve problems by definition! Perhaps even in the case of a prime ring R it might be impossible to show that R itself or various important subsets of R were d-free for any d.Fortunately Beidar and Chebotar took care in [7] to make sure this was not the case. The notion of *degree* of a prime ring R over C is a familiar one: $\deg(R)$ is the maximum degree (possibly ∞) of its elements over C. In [7, Theorem 2.4] Beidar and Chebotar noted that in essence Beidar and I [12] had shown (i) for R prime $\deg(R) \geq d$ implies R is d-free and (ii) for R prime with involution $\deg(R) \geq 2d + 1$ implies K is d-free. More generally it has been shown that (iii) for R prime with involution and U a non-central Lie ideal of K (e.g. $[K, K]$) then $\deg(R) \geq 2d + 3$ implies U is d-free.

In Theorem 4 there was the advantage of having the domain closed under squares, and even in Theorem 5 there was a similar advantage of having the domain closed under cubes. But what about Lie isomorphisms of $[A, A]$ or of the "derived" Lie ring of skew elements $[L, L]$? Up to this point all results (e.g. [20]) in these situations had been obtained under the assumption of idempotents. Beidar and Chebotar were to make a major breakthrough [9, Lemma 2.5] in this matter by proving the following

Lemma 3. *If S is a Lie ideal of a ring A, $\alpha : S \to Q/C$ is a Lie homomorphism, C is a direct additive summand of Q, and $S^\alpha = T/(T \cap C)$ where T is 7-free in Q, then α can be extended to a Lie homomorphism $\gamma : U \to Q/C$, with U a Lie ideal of A containing S and closed under squares.*

The proof of this lemma required the use of [8, Theorem 2.11], some heavy machinery to say the least. Beidar and Chebotar went on to verify Herstein's Lie homomorphism conjectures [9, Theorem 1.2] and in a subsequent paper [10, Theorem 1.2] Herstein's Lie derivation conjectures, both in the "non-involution" situation and modulo "low-dimensional" cases.

By now the four of us (Beidar, Brešar, Chebotar, and myself) had decided to join forces and dedicate ourselves to the complete resolution of Herstein's Lie map conjectures [3, 4, 5]. In view of the result [9] of Beidar and Chebotar just described we naturally turned our attention to the Lie homomorphism problem in the more involved involution case. The idea of how to proceed, of course, had already been given in Lemma 3. I remember being an admiring bystander as my three colleagues proceeded to prove (and then make use of) the analogue [3, Lemma 3.2] of Lemma 3: given a ring A, a Lie subring L of A, a Lie ideal S of L, and a Lie homomorphism $\alpha : S \to Q/C$ then (under a suitable d-freeness assumption) α may be extended to a Lie homomorphism $\gamma : V \to Q/C$, where V is a Lie ideal of L such that V is closed under cubes. Here is one of the theorems of our paper [3, Theorem 3.5] which essentially solves all of Herstein's Lie isomorphism problems in the involution case. It is to be noted that the surprising generality of this result is largely due to the machinery developed by Beidar and Chebotar in [8] and [9].

Theorem 6. *Let A be a ring, let L be a Lie subring of A with $L \cap (L \circ L) = 0$ and $xyz + zyx \in L$ for all $x, y, z \in L$, and let V be a Lie ideal of L. Let Q be a unital ring with center C as an additive direct summand, and let U be an additive subgroup of Q which is 9-free in Q. If $\alpha : V \to U/(U \cap C)$ is a surjective Lie homomorphism, then there exists a homomorphism $\sigma : \langle V \rangle \to \langle U \rangle C + C$ such that $x^\alpha = \overline{x^\sigma}$ for all $x \in V$.*

Of course this result is a thinly disguised way of saying what the authors really have in mind: A is a ring with involution, L is the skew elements of A, V might easily be $[L, L]$, Q is the maximal right ring of quotients of a prime ring R with involution and with skew elements K, U might well be $[K, K]$, and $\deg(R) \geq 21$ ($= 2(9) + 3$ in view of (iii) above).

Now I want to touch again on the analogous problems for Lie derivations. As I mentioned in section 1 derivation proofs seem inherently easier than homomorphism proofs. On the other hands derivation proofs are apt to be longer (just try applying a derivation to a product of m elements). This is born out in Swain's paper [33] in which he obtained the analogue of Theorem 5 for Lie derivations (see also the paper [34] of Swain and Blau for handling the situation when the involution is of the second kind).

In view of observation (1) made in section 1 it is tempting to hope to get theorems about derivations "free of charge" from the corresponding homomorphism theorems. The catch, of course, is that the image of the induced homomorphism via (1) may no longer be a suitable one. But, thanks to an idea of Brešar, we have the following rather general d-freeness result [4, Theorem 2.1]:

Theorem 7. *Let S be a d-free subset of a unital ring Q and let $\widehat{Q}$ denote the ring of all matrices of the form $\begin{pmatrix} x & y \\ 0 & x \end{pmatrix}$ where $x, y \in Q$. If $\delta : S \to Q$ is any map whatsoever, then the image of the map given by (1) is a d-free subset of $\widehat{Q}$.*

Now let's say we are faced with a Lie derivation δ of the skew elements K of a prime ring R with involution and with the hypothesis that $\deg(R) \geq 19$. Then we already know (from (ii) above) that K is a 9-free subset of Q. Clearly the map $\alpha : K \to \widehat{Q}$ given by (1) is a Lie homomorphism. By Theorem K^α is a 9-free subset of $\widehat{Q}$. By Theorem 6 there exists a homomorphism $\sigma : \langle K \rangle \to \widehat{Q}$ such that $x^\sigma = x^\alpha + \mu(x)$ for all $x \in K$, where $\mu(x) \in \widehat{C}$ the center of $\widehat{Q}$. Clearly $x^\sigma = x^\alpha$

for all $x \in [K,K]$, from which it is easy to show that for all $x \in \langle [K,K] \rangle$ we have $x^\sigma = \begin{pmatrix} x & x^d \\ 0 & x \end{pmatrix}$. It follows that d must be a derivation of $\langle [K,K] \rangle$ with $x^d = x^\delta$ for all $x \in [K,K]$. More can be said, to be sure, but for our purposes this already illustrates the power of some of the results on d-freeness. In particular it shows, in the case of Theorem 6 above, the usefulness of having the more general assumption of Q being an arbitrary unital ring rather than just taken as the maximal right ring of quotients of a prime ring.

The downside of d-freeness as a tool in proving the various Lie isomorphism theorems is that it can only be used when the degree of the prime ring R in question is sufficiently high (in fact, in view of (iii) all the theorems will hold if $\deg(R) \geq 21$). However, when $\deg(R) < 21$, we are in essence in a world teeming with idempotents, in which case Herstein's problems had for the most part already been solved. The complete details have been written up in [5]. With one isolated exception, counterexamples (the worst one occurring in the 8×8 matrices) have been found illustrating all restrictions assumed in the various theorems; a list of these is given in [5, section 5]. Needless to say, I was privately happy to see that arguments involving idempotents and Pierce decompositions (to which I had devoted so much energy) still had their place in the complete proofs of Herstein's Lie map problems.

The study of Lie homomorphisms suggests that it might be of interest to study in more generality additive mappings $\alpha : S \to Q$ which preserve a multilinear polynomial $f(x_1, x_2, \ldots, x_m) \in Z\langle X \rangle$(see [6] for references). Such a mapping will be referred to as an f-homomorphism. Jordan homomorphisms have been studied extensively. Herstein proved that if A is a unital ring, Q is a prime ring of characteristic 0 or greater than m, and $\alpha : A \to Q$ is an additive surjection such that $(x^m)^\alpha = (x^\alpha)^m$ for all $x \in A$, then there exists $\lambda \in C$ such that $\lambda^{m-1} = 1$ and $x^\alpha = \lambda x^\sigma$, where σ is an (anti-)homomorphism of A onto Q (this situation can be made multilinear by linearizing x^m by the usual process). Note that the conclusion concerning λ is to be expected since the map $x \to \lambda x$ preserves mth powers. A few other results for $m > 2$ had also been obtained.

Beidar and Fong, in their 1999 paper [11], were the first to give a general and unified treatment of this subject (for f-homomorphisms of a prime ring A onto a prime ring Q). Naturally this suggested looking for a general theorem in the involution context, and what our group proved [6, Theorem 1.1] was the following (simplified a bit):

Theorem 8. *Let A be an F-algebra with involution (F a field) with symmetric elements S, and let f be a multilinear polynomial of degree m over F such that S is closed under f. Next let R be a prime F-algebra with involution of degree $> \max\{6m + 1, 15\}$ over C, with symmetric elements T, extended center C, and maximal right ring of quotients Q. Let $\alpha : S \to Q$ be an f-homomorphism such that $S^\alpha \supseteq T$ If either f or α satisfies some rather general conditions (which I won't mention here), then there exists $\lambda \in C$, with $\lambda^{m-1} = 1$, a homomorphism $\sigma : \langle S \rangle \to Q$, and a linear map $\tau : S \to C$ such that $x^\alpha = \lambda x^\sigma + \tau(x)$ for all $x \in S$.*

This theorem is a bit deceptive, since it is really three separate results rolled into one if one examines the proof. First it is shown that $[[x, y], z]^\alpha$ satisfies one of the heavy-duty FI's dealt with in [8] with the result that $[[x, y], z]^\alpha = \lambda[[x^\alpha, y^\alpha], z^\alpha]$ (modulo C). Thus we have already descended from the stratosphere almost down to earth! Next, since $[[x, y], z] = (z \circ y) \circ x - (z \circ x) \circ y$ it is not surprising that one can then show that $(x \circ y)^\alpha = \lambda(x^\alpha \circ y^\alpha)$ (modulo C). Therefore we are essentially able to assume that α is a Jordan homomorphism. I would like to dwell a bit on this case, since part of the argument makes crucial use of one of the Lie homomorphism theorems discussed earlier. First, without loss of generality we may assume that $U = S \oplus [S, S]$ is a Lie ideal of A [6, Lemma 3.8]. It is then natural to define $\gamma : U \to Q$ according to $x + \sum[y_i, z_i] \to x^\alpha + \sum[y_i^\alpha, z_i^\alpha]$ and then to show that γ is a Lie homomorphism. By [9, Theorem 2.7], γ (and hence α) is induced by $\sigma : \langle U \rangle \to Q$, where σ is either a homomorphism or the negative of an anti-homomorphism. This latter case is then ruled out, which completes the proof.

The degree condition on R appears because it ensures that various key subsets are d-free for sufficiently high d. As an extreme example suppose that both A and R satisfy the standard polynomial $f = S_n$. Then any map whatsoever from A into Q is trivially an f-homomorphism. So the degree of R over C must be sufficiently high in order to avoid this pitfall. Unlike the situation with Lie homomorphisms, in which either proofs or counterexamples have been given for all cases not covered by d-freeness arguments, i.e., the "low-dimensional" cases, I could well imagine someone spending a lifetime (or several lifetimes!) trying to work out proofs and counterexamples for Theorem 8 if the degree of R over C is less than $\max\{6m + 1, 15\}$.

In conclusion I believe that both of my colleagues Brešar and Chebotar would concur with me that Beidar was the leading force of our group in finally settling (in the 1990's) all of the Lie map conjectures given by Herstein in his 1961 AMS talk.

References

[1] Amitsur, S. A. *Generalized polynomial identities and pivotal monomials*, Trans. A. M. S. **114** (1965), 210–226.

[2] Ayupov, Sh. A. *Skew commutators and Lie isomorphisms in real von Neumann algebras*, J. Funct. Anal. **138** (1996), 170–187.

[3] Beidar, K. I.; Brešar, M.; Chebotar, M. A.; Martindale, W. S., 3rd, *On Herstein's Lie map conjectures I*, Trans. A. M. S. **353** (2001), 4235–4260.

[4] Beidar, K. I.; Brešar, M.; Chebotar, M. A.; Martindale, W. S., 3rd, *On Herstein's Lie map conjectures II*, J. Algebra **238**(2001), 239–264.

[5] Beidar, K. I.; Brešar, M.; Chebotar, M. A.; Martindale, W. S., 3rd, *On Herstein's Lie map conjectures III*, J. Algebra **249**(2002), 59–94.

[6] Beidar, K. I.; Chebotar, M. A.; Martindale, W. S., 3rd, *Polynomial preserving maps on certain Jordan algebras*, Israel J. Math. **141** (2004), 285–313.

[7] Beidar, K. I.; Chebotar, M. A. *On functional identities and d-free subsets of rings I*, Comm. Algebra **28** (2000), 3925–3951.

[8] Beidar, K. I.; Chebotar, M. A. *On functional identities and d-free subsets of rings II*, Comm. Algebra **28** (2000), 3953–3972.

[9] Beidar, K. I.; Chebotar, M. A. *On surjective Lie homomorphisms onto Lie ideals of prime rings*, Comm. Algebra **29** (2001), 4775–4793.

[10] Beidar, K. I.; Chebotar, M. A. *On Lie derivations of Lie ideals of prime algebras*, Israel J. Math. **123** (2001), 131–148.

[11] Beidar, K. I.; Fong, Y. *On additive isomorphisms of prime rings preserving polynomials*, J. Algebra **217** (1999), 650–667.

[12] Beidar, K. I.; Martindale, W. S., 3rd. *On functional identities in prime rings with involution*, J. Algebra **203** (1998), 491–532.

[13] Beidar, K. I.; Martindale, W. S., 3rd; Mikhalev, A. V. *Lie isomorphisms in prime rings with involution*, J. Algebra **169** (1994), 304–327.

[14] Beidar, K. I.; Martindale, W. S., 3rd; Mikhalev, A. V. *Rings with generalized identities*, Dekker, New York, 1996.

[15] Blau, P. S. *Lie isomorphisms of prime rings satisfying S_4*, Southeast Asian Math. Bull. **25** (2002), 581–587.

[16] Brešar, M. *Commuting traces of biadditive mappings, commutativity preserving mappings, and Lie mappings*, Trans. A. M. S. **335** (1993), 525–546.

[17] Chebotar, M. A. *On Lie isomorphisms in prime rings with involution*, Comm. Algebra **27** (1999), 2767–2777.

[18] Erickson, T. S. *The Lie structure in prime rings with involution*, J. Algebra **21** (1972), 523–534.

[19] Herstein, I. N. *Lie and Jordan structures in simple, associative rings*, Bull. A. M. S. **67** (1961), 517–531.

[20] Howland, R. A. *Lie isomorphisms of derived rings of simple rings*, Trans. A.M.S. **145** (1969), 383–396.

[21] Hua, L. *A theorem on matrices over an sfield and its applications*, J. Chinese Math. Soc. (N. S.) **1** (1951), 110–163.

[22] Jacobson, N. Lie algebras, Interscience tracts in pure and applied math., no. 10, Interscience, New York, 1962.

[23] Lambek, J. Lectures on rings and modules, Blaisdell, Waltham MA, 1966.

[24] Martindale, W. S., 3rd, *Lie isomorphisms of primitive rings*, Proc. A. M. S. **14** (1963), 909–916.

[25] Martindale, W. S., 3rd, *Lie derivations of primitive rings*, Mich. Math. J. **11** (1964), 183–187.

[26] Martindale, W. S., 3rd, *Lie isomorphisms of prime rings*, Trans. A. M. S. **142** (1969), 437–455.

[27] Martindale, W. S., 3rd, *Prime rings satisfying a generalized polynomial identity*, J. Algebra **12** (1969), 576–584.

[28] Martindale, W. S., 3rd, *Prime rings with involution and generalized polynomial identities*, J. Algebra **22** (1972), 502–516.

[29] Martindale, W. S., 3rd *Lie isomorphisms of the skew elements of a prime ring with involution*, Comm. Algebra **4**(1976), 929–977.

[30] C.R. Miers, *Lie iosmorphisms of factors*, Trans. A. M. S. **147** (1970), 55–63.

[31] Miers, C. R. *Lie derivations of von Neumann algebras*, Duke Math. J. **40** (1973), 403–409.

[32] Rosen, M. P. *Isomorphisms of a certain class of prime Lie rings*, J. Algebra **89** (1984), 291–317.

[33] Swain, G. A. *Lie derivations of the skew elements of prime rings with involution*, J. Algebra **184** (1996), 679–704.

[34] Swain, G. A.; Blau, P. S. *Lie derivations in prime rings with involution*, Canad. Math. Bull. **42** (1999), 401–411.

Wallace S. Martindale, 3rd, Department of Mathematics, University of Massachusetts, Amherst, USA

E-mail address: jerrywsm@comcast.net

Our joint work with Kostia Beidar on the separativity problem for regular rings

K. C. O'Meara and R. M. Raphael

We dedicate this article to the Beidar family, our friend and collaborator Kostia, his wife Lydia, and their wonderful daughters Katya and Tanya. We cherish the time that our families spent together in New Zealand.

1. Introduction

Fate ruled that we would write only one paper with Kostia Beidar, namely [4]. More work on the separativity problem had been discussed. Below we explain the context of our work and some of the directions it took. We wish to thank Wen-Fong Ke and Misha Chebotar for their work in preparing this commemorative volume for Kostia.

2. Separativity and regular rings

In what follows all algebras contain an identity. It is assumed that any subalgebra contains the identity of the algebra. The concept of separativity has played a unifying role for certain direct sum cancellation problems in ring theory, C^*-algebras and abelian group theory [1, 2, 3, 5, 14, 16, 17]. Within the class of (von Neumann) regular rings, it has been the unifying theme for a number of outstanding open problems [2].

Recall that a (von Neumann) regular ring R is said to be *separative* if for all finitely generated projective R-modules A and B, the conditions $A \oplus A \cong A \oplus B \cong B \oplus B$ imply $A \cong B$. The fundamental Separativity Problem [2] for regular rings asks whether such rings are always separative. A positive answer to this question would simultaneously give a positive answer to five outstanding open problems in regular rings R.

(1) If R is directly finite (one sided inverses are two-sided), is the same true of its matrix rings $M_n(R)$? (If so, a finitely generated projective module is not isomorphic to a proper direct summand of itself).

(2) If R is simple and directly finite, does R have stable range 1? (If so, a finitely generated projective module could always be cancelled from a direct sum).

(3) Are $1, 2$ and ∞ the only possible stable range values of R?

(4) Is every square matrix A over R equivalent to a diagonal matrix, i.e. is PAQ diagonal for some invertible matrices P and Q?

(5) Is every invertible square matrix over R a product of elementary matrices and an invertible diagonal matrix? (If so, then the natural homomorphism from $GL_1(R)$ to the Whitehead group $K_1(R)$ is surjective).

All of these questions have been shown to have a positive answer within the class of separative regular rings. (See references [1, 2, 3].) Moreover, separativity itself is a very robust property, being preserved in all standard constructions (except perhaps inverse limits). In particular, it is preserved under extensions of ideals by factor rings, which makes constructions of non-separative regular rings difficult. Despite all this, the "smart money" says non-separative regular rings should exist. (If non-separative regular rings do exist, there will be countable ones as well.)

Our work with Kostia was based on an equivalence between separativity for regular rings and a certain linear algebra problem established a few years earlier in [13].

In order to state the equivalence, we need the following definition.

Definition 2.1. Let Λ be a commutative ring and let R be a regular Λ-algebra. As in [13] we say that 2×2 matrices over R can be *uniformly diagonalised* by regular Λ-algebra operations if for a general 2×2 matrix

$$A = \begin{pmatrix} a & b \\ c & d \end{pmatrix}$$

over R, there exist fixed regular Λ-algebra expressions $\alpha_1, \ldots, \alpha_4, \beta_1, \ldots, \beta_4$ in a, b, c, d such that for *all* substitutions for a, b, c, d from R, and for *all* choices of a quasi-inverse operation $'$ in R (recall that this means $xx'x = x$ for all $x \in R$), the matrices

$$P = \begin{pmatrix} \alpha_1 & \alpha_2 \\ \alpha_3 & \alpha_4 \end{pmatrix} \quad \text{and} \quad Q = \begin{pmatrix} \beta_1 & \beta_2 \\ \beta_3 & \beta_4 \end{pmatrix}$$

are invertible and

$$PAQ = \begin{pmatrix} * & 0 \\ 0 & * \end{pmatrix}$$

is diagonal.

The following result was key to our work.

Theorem 2.2 ([13, Theorem 2.5]). *Let F be a field. Then all regular F-algebras are separative if and only if 2×2 matrices over $M_n(F)$ can be uniformly diagonalised* independently *of n.*

We now set in place some further notation. Let T be the free regular Λ-algebra on the set $\{x_1, x_2, x_3, x_4\}$ [9]. We define subalgebras T_i of T inductively as follows: T_0 is the subalgebra of T generated by $\{x_1, x_2, x_3, x_4\}$; T_{i+1} is the subalgebra of T generated by T_i and the set $\{a' \mid a \in T_i\}$, $i = 0, 1, \ldots$. Clearly $T = \cup_{i=0}^{\infty} T_i$. Given $f \in T$, we set

$$L(f) = \min\{i \mid f \in T_i\}.$$

Next, let S be the subalgebra of T generated by $\{x_i, x_i' \mid 1 \leq i \leq 4\}$.

By the *diagonalisation complexity* of a regular ring R one means the least k for which T_k contains all the entries of the matrices P and Q in some uniform diagonalisation formula for 2×2 matrices over R (or ∞ if there is no uniform diagonalisation formula). Thus 'complexity' records the minimum number of layered (or recursive) uses of the quasi-inverse operation $'$ in a uniform diagonalisation formula, not the actual number of occurrences of $'$. For example, the expression

$$(((1 + ab + c^2 d)' + b^5)')^7 + c' + (d'a)^4$$

has complexity 2. Theorem 3.3 below shows that m-rings are precisely the regular rings of diagonalisation complexity 0, and Theorem 1.4 shows that abelian regular rings have complexity at most 1 (although possibly there are others as well).

Let $p \in T$, let R be a regular Λ-algebra with quasi-inverse operation $' : R \to R$, let $a_1, a_2, a_3, a_4 \in R$ and let $\phi : T \to R$ be a homomorphism of regular algebras such that $\phi(x_i) = a_i$ for each i. We shall write $p(\bar{a})$ (or $p(a_1, a_2, a_3, a_4)$) for $\phi(p)$. It will always be assumed that $0' = 0$.

It is widely believed that the fundamental Separativity Problem for regular rings has a negative answer (see [2, Section 6]). Theorem 2.2 suggests a possible way to show this: prove that there is no uniform formula for diagonalising 2×2 matrices over $M_n(F)$ (here F is a field) which is independent of n. For example, one could try to show that there is no finite upper bound (independent of n) on the diagonalisation complexity of the rings $M_n(F)$.

This approach was initiated in [13] and the following result was obtained.

Theorem 2.3 ([13, Theorem 3.10]). *Let R be a Λ-algebra. Then the following are equivalent:*

(i) 2×2 *matrices over R can be uniformly diagonalised by Λ-algebra operations alone (i.e., there exist P and Q as in Definition 2.1 with $\alpha_i, \beta_i \in T_0$);*
(ii) *there exists an integer m with $m > 1$ such that $x^m = x$ for all $x \in R$ (i.e.,
R is an m-ring);*
(iii) *R is unit regular and every element $x \in R$ has a unit quasi-inverse that is
given by a (fixed) polynomial in x over Λ;*
(iv) *R is regular and every element $x \in R$ has a quasi-inverse that is given by
a (fixed) polynomial in x over Λ.*

In our work with Kostia we succeeded in making the next step within this approach by showing the following.

Theorem 2.4. *Let Λ be a commutative ring, let T be the free regular Λ-algebra
on the set $\{x_1, x_2, x_3, x_4\}$ and let S be the subalgebra of T generated by the set
$\{x_i, x_i' \mid 1 \leq i \leq 4\}$. Further, let R be a regular Λ-algebra. Then the following
conditions are equivalent:*

(a) *there exist $p_i, q_i \in S$, $1 \leq i \leq 4$, such that for any $a_1, a_2, a_3, a_4 \in R$ and
any choice of the quasi-inverse operation $'$ on R both of the matrices*

$$P(\bar{a}) = \begin{pmatrix} p_1(\bar{a}) & p_2(\bar{a}) \\ p_3(\bar{a}) & p_4(\bar{a}) \end{pmatrix} \quad and \quad Q(\bar{a}) = \begin{pmatrix} q_1(\bar{a}) & q_2(\bar{a}) \\ q_3(\bar{a}) & q_4(\bar{a}) \end{pmatrix}$$

are invertible and the matrix

$$P(\bar{a}) \begin{pmatrix} a_1 & a_2 \\ a_3 & a_4 \end{pmatrix} Q(\bar{a})$$

is diagonal;
(b) *R is abelian regular (i.e., all idempotents of R are central, equivalently, R
has no nonzero nilpotents).*

In the intervening years no progress seems to have been made in this vein and
the separativity question for regular rings remains open.

3. One-accessible regular algebras

If general separativity holds then the universal diagonalisation formula will have
to be expressible by finitely many words from the free regular algebra in four
variables and these will contain only a finite number of occurrences of the quasi-
inverse operation. Thus, tangentially, separativity raised the question of how many
applications of the quasi-inverse operation are needed to make a subalgebra regular. In the case of m-rings [13] the answer is really zero because one gets the

quasi-inverse of an element as a power. In the general commutative regular case, it follows from work by Olivier [12] that the adjunction of a single quasi-inverse for each element of the subalgebra makes the subalgebra regular.

This motivated the following notion.

Definition 3.1. Let R be a regular algebra over a field F. We say that R is $1 - ac$ (one-accessible) if for any $a \in R$ there exist $b \in F[a]$ and $b' \in R$ such that $bb'b = b$, $b'bb' = b'$ (such elements b' are called g-inverses of b [13]) and the subalgebra $F[a, b']$ of R, generated by a and b', is regular.

Here are some of the results that were obtained.

Proposition 3.2. *Let R be a regular algebra over a field F.*

  (i) *If a division algebra over F is $1 - ac$, then it is algebraic.*
 (ii) *A homomorphic image of a regular $1 - ac$ algebra is again $1 - ac$.*
(iii) *If R is a commutative regular F-algebra, then R is $1 - ac$ iff it is an algebraic algebra.*
 (iv) *The center of a regular $1 - ac$ F-algebra is $1 - ac$.*
  (v) *An infinite direct product of regular $1 - ac$ F-algebras need not be $1 - ac$. Thus regular $1 - ac$ algebras do not form a variety.*
 (vi) *The free regular F-algebra [9] is not $1 - ac$.*
(vii) *In any regular F-algebra there exists (by Zorn's lemma) a maximal $1 - ac$ regular F-subalgebra.*

For the main theorem one needs to recall that a field F is called *perfect* if any finite field extension of F is separable.

Theorem 3.3. *Let R be a regular algebra over a field F. Then:*

  (i) *if F is uncountable and R is $1 - ac$, then R is algebraic;*
 (ii) *a finite product of algebraic central simple Artinian F-algebras is $1 - ac$; for example this holds for a finite product of central simple finite dimensional F-algebras ;*
(iii) *a regular algebraic algebra over a perfect (e.g. algebraically closed) field is $1 - ac$;*
 (iv) *if F is an uncountable perfect (e.g. algebraically closed) field, then R is $1 - ac$ if and only if R is algebraic.*

One particular step needed in the proof of this theorem took on a life of its own. A close inspection of [8, Lemma 7.1], established by K. R. Goodearl in the 1970's, showed that one could replace the regularity of the ring in question by the weaker condition of the regularity of all powers of the considered nilpotent element. Thus one had.

Lemma 3.4. *Let R be any ring, let A be a projective right R-module and let $t \in \mathrm{End}_R(A)$. Suppose that t is a nilpotent endomorphism of index n and all powers of t are regular elements of $\mathrm{End}_R(A)$. Then there exists a decomposition $A = A_1 \oplus A_2 \oplus \ldots \oplus A_n$ such that $tA_i = A_{i+1}$ for $i = 1, 2, \ldots, n-1$ and $tA_n = 0$.*

The following result was obtained. It implies the theorem in [6, p. 370], and Yu's unit-regularity result for nilpotents with regular powers in strongly π-regular rings [18].

Theorem 3.5. *Let R be an algebra over a commutative ring Λ and let $a \in R$ be a nilpotent element of index n. Then the following are equivalent:*

    (i) *each power of a is regular;*

    (ii) *each power of a is unit-regular;*

    (iii) *there exist an element $b \in R$, integers $1 \leq n_1 < n_2 < \ldots < n_\ell = n$ and ideals $I_1, I_2, \ldots, I_\ell$ of the ring Λ such that $a = aba$, $bab = b$, $\cap_{j=1}^{\ell} I_j = \mathrm{ann}_\Lambda(R)$, the algebra $\Lambda[a, b]$ is isomorphic to $\prod_{j=1}^{\ell} M_{n_j}(\Lambda/I_j)$, and for each j the canonical image of a in $M_{n_j}(\Lambda/I_j)$ is the $n_j \times n_j$ matrix*

$$a_j = \begin{bmatrix} 0 & & & & \\ 1 & 0 & & & \\ & 1 & \ddots & & \\ & & \ddots & 0 & \\ & & & 1 & 0 \end{bmatrix}.$$

If, in addition, Λ is regular, then (i)–(iii) are equivalent to:

    (iv) *a lies in a unit-regular Λ-subalgebra of R.*

4. A new canonical form for matrices

In an interesting sequence of events, Lemma 3.4 and the proof of Theorem 3.5 have turned out to provide exactly the right ring-theoretic backdrop for a new canonical form, the H-form, for matrices over an algebraically closed field F. ("H" stands for "Husky", in recognition of the University of Connecticut connection.) The H-form is introduced by the first author and C. Vinsonhaler in [15], as a tool for tackling the following question which arose in a recent study of phylogenetic invariants in biomathematics: Given $A_1, A_2, \ldots, A_k$ commuting $n \times n$ matrices over the complex numbers $\mathbb{C}$, can the matrices be perturbed by an arbitrarily small amount so that they become simultaneously diagonalizable? (The case $k = 2$ was answered positively in 1955 by Motzkin and Taussky [11], Theorem 5. A related classical problem has been studied by a number of people, starting with Gerstenhaber [7] in 1961; that of bounding the dimension over $\mathbb{C}$ of $\mathbb{C}[A_1, A_2, \dots, A_k]$, the subalgebra (with identity) of the $n \times n$ complex matrices generated by commuting $A_1, A_2, \dots, A_k$.) Informally, a matrix A is in *H-form* if A is a direct sum of basic H-matrices, one for each distinct eigenvalue of A, and a *basic H-matrix* is a blocked-matrix generalization of a basic Jordan matrix

$$\begin{bmatrix} \lambda & 1 & & & \\ & \lambda & 1 & & \\ & & \ddots & \ddots & \\ & & & \lambda & 1 \\ & & & & \lambda \end{bmatrix}$$

with associated eigenvalue λ. In the blocked form, we replace the λ's by scalar matrices λI for various identity matrices I in decreasing order of size going down the diagonal blocks. The 1's are replaced by full column rank matrices in reduced row echelon form, that is

$$\begin{bmatrix} I \\ 0 \end{bmatrix}$$

an identity matrix followed by zero rows. In the blocked form, these matrices make up the first super-diagonal of blocks, and their sizes are dictated by the diagonal block sizes. All other blocks are zero. For example,

$$\left[\begin{array}{cc|cc|c|c|c} \lambda & 0 & 1 & 0 & & & \\ & \lambda & 0 & 1 & & & \\ \hline & & \lambda & 0 & 1 & & \\ & & & \lambda & 0 & & \\ \hline & & & & \lambda & 1 & \\ \hline & & & & & \lambda & 1 \\ \hline & & & & & & \lambda \end{array}\right] \quad \text{and} \quad \left[\begin{array}{ccc|ccc} \lambda & 0 & 0 & 1 & 0 & 0 \\ & \lambda & 0 & 0 & 1 & 0 \\ & & \lambda & 0 & 0 & 1 \\ \hline & & & \lambda & 0 & 0 \\ & & & & \lambda & 0 \\ & & & & & \lambda \end{array}\right]$$

are legitimate basic H-matrices with block structures $(2, 2, 1, 1, 1)$ and $(3, 3)$ respectively. The H-form of a matrix is unique. And although the H-form is conjugate to the Jordan form under a permutation transformation (the H-structure and Jordan structure of an $n \times n$ nilpotent matrix are actually conjugate partitions of n), the H-form seems better suited for studying commuting relationships. For example, given a finite list of commuting matrices, one can simultaneously put the first one in H-form and the rest in upper triangular form, something not possible with the Jordan form. Another interesting contrast is in the description of the centralizer of a nilpotent matrix in terms of its H-structure $(n_1, n_2, \ldots, n_r)$ and Jordan structure $(m_1, m_2, \ldots, m_s)$. With the former the description is quite simple (similar to what happens with a basic Jordan matrix) and

$$\dim \mathcal{C}(A) = n_1^2 + n_2^2 + \cdots + n_r^2.$$

With the latter, the description is more complicated and

$$\dim \mathcal{C}(A) = m_1 + 3m_2 + 5m_3 + \cdots + (2s - 1)m_s.$$

(Gerstenhaber observed the equality of these two formulae in [7] Proposition 8, using induction and conjugate partitions, but did not make the matrix structure connection with the first.)

Having said what the H-form of a matrix is, we now comment on its connection with our joint work with Kostia. The following module-theoretic formulation of an H-form suggests that the H-form lives in a somewhat bigger universe than its Jordan counterpart, even though each can be derived from the other for matrices over an algebraically closed field. (Even in that special setting, the H-form may be a little more "basis-free".)

Definition 4.1. Suppose $\tau : P \to P$ is a nilpotent endomorphism of a (nonzero) projective module P over an arbitrary (and noncommutative) ring R. Then an H-form for τ is a direct sum decomposition

$$P = P_1 \oplus P_2 \oplus \cdots \oplus P_r$$

of P into nonzero submodules such that τ annihilates P_1 and maps P_i isomorphically onto a direct summand of P_{i-1} for $i = 2, \ldots, r$.

This brings us back to Lemma 3.4 and Theorem 3.5, which can be used to establish the following condition for the existence of an H-form.

Proposition 4.2. *Let $\tau : P \to P$ be a nilpotent endomorphism of a projective module P over a ring R. Then τ has an H-form precisely when all the powers τ^k of τ are (von Neumann) regular in the endomorphism ring $\mathrm{End}_R(P)$. (Recall that an element a of a ring S is* regular *if $a = aba$ for some $b \in S$, equivalently, aS is a direct summand of S.)*

Of course, when $\tau : V \to V$ is a nilpotent linear transformation of a finite-dimensional vector space V over any field F, Proposition 4.2 applies (taking $R = F$ and $P = V$) so τ has an H-form as a transformation. If the H-decomposition of V is $V = V_1 \oplus V_2 \oplus \cdots \oplus V_r$, one then obtains an H-form for a matrix of τ in a basis $\mathcal{B} = \mathcal{B}_1 \cup \mathcal{B}_2 \cup \cdots \cup \mathcal{B}_r$ constructed as follows. Start with any basis $\mathcal{B}_r$ for V_r. Next extend $\tau(\mathcal{B}_r)$ to a basis for V_{r-1} and call this $\mathcal{B}_{r-1}$. Continue in this way to inductively construct the $\mathcal{B}_i$ for $i = r, r - 1, \ldots, 2$. Finally, take $\mathcal{B}_1$ to be any basis for V_1 (which is the null space of τ).

References

[1] Ara, P.; Goodearl, K. R.; O'Meara K. C.; Pardo, E. *Diagonalisation of matrices over regular rings*, Linear Algebra Appl. **265** (1997), 147–163.

[2] Ara, P.; Goodearl, K. R.; O'Meara K. C.; Pardo, E. *Separative cancellation for projective modules over exchange rings*, Israel J. Math. **105** (1998), 105–137.

[3] Ara, P.; Goodearl, K. R.; O'Meara K. C.; Raphael, R. M. *K_1 of separative exchange rings and C^*-algebras with real rank zero*, Pacific J. Math. **195** (2000), 261–275.

[4] Beidar, K. I.; O'Meara, K. C.; Raphael, R. M. *On Uniform Diagonalistion of Matrices over Regular Rings and One-Accessible Regular Algebras*, Comm. Algebra **32** (2004), 3543–3562.

[5] Brookfield, G. *Direct sum cancellation of noetherian modules*, J. Algebra **200** (1998), 207–224.

[6] Farkas, D. R.; Snider, R. L. *Locally finite-dimensional algebras*, Proc. Amer. Math. Soc. **81** (1981), 369–372.

[7] Gerstenhaber, M. *On dominance and varieties of commuting matrices*, Ann. of Math. **73** (1961), 324–348.

[8] Goodearl, K. R. Von Neumann Regular Rings, Pitman, London, 1979; Second Ed., Krieger, Malabar, Fl. 1991.

[9] Goodearl, K. R.; Menal, P.; Moncasi, J. *Free and residually artinian regular rings*, J. Algebra **156** (1993), 407–432.

[10] Jacobson, N. Basic Algebra II, Freeman, 1980.

[11] Motzkin, T.; Taussky, O. *Pairs of matrices with propery L II,* Trans. Amer. Math. Soc. **80** (1955), 387–401.

[12] Olivier, J. P. *Anneaux absolument plats universels et épimorphismes d'anneaux,* C. R. Acad. Sci. Paris Seŕ. A-B, **266** (1968), 317–318.

[13] O'Meara, K. C.; Raphael, R. M. *Uniform diagonalisation of matrices over regular rings*, Algebra Universalis **45** (2001), 383–405.

[14] O'Meara, K. C.; Vinsonhaler, C. *Separative cancellation and multi-isomorphism in torsion-free abelian groups*, J. Algebra **221** (1999), 536–550.

[15] O'Meara, K. C.; Vinsonhaler, C. *On approximately simultaneously diagonalizable matrices*, Linear Algebra Appl. **412** (2006), 39–74.

[16] Pardo, E. *Comparability, separativity, and exchange rings*, Comm. Algebra **24** (1996), 2915–2929.

[17] F. Perera, *Lifting units modulo exchange ideals and C^*-algebras with real rank zero*, J. Reine Angew. Math. **522** (2000), 51–62.

[18] Yu, H. P. *On strongly π-regular rings of stable range one*, Bull. Austral. Math. Soc. **51** (1995), 433–437.

K. C. O'Meara, Department of Mathematics, University of Connecticut, Storrs CT 06269, USA
E-mail address: `staf198@ext.canterbury.ac.nz`

R. M. Raphael, Department of Mathematics, Concordia University, Montreal, Canada H4B 1R6.
E-mail address: `raphael@alcor.concordia.ca`

Kostia's contribution to radical theory and related topics

E. R. Puczyłowski and R. Wiegandt

Abstract. One of Kostia Beidar's many research fields was the theory of radicals, concrete and general ones, and the structure of rings which are radical-free relative to a certain radical or which arise from certain radical rings. His rich contribution to this area and its impact to further researches were considerable, substantial and important. We intend to give an overview on a segment of his contribution to algebra, and shall address the main points of this topic.

Contents

Mathematics Subject Classification 2000: 16N20, 16N40, 16N60, 16N80, 16W10, 17A60, 17A65, 17D05.

The authors gratefully acknowledge the support of the National Cheng Kung University, Taiwan ROC., the Polish KBN grant No. 1 P03A 032 27 and the Hungarian OTKA grant T043034.

Introduction

One of Kostia Beidar's main research interest was the theory of radicals and the description of the structure of rings which are radical or radical-free with respect to a certain radical. Kostia's research activity ranges over his whole scientific career, he published radical theoretical results instantaneously since 1981 till his untimely death in some 30 papers. In 1981-1982 Kostia has become famous in ring-theoretical circles by solving the Amitsur Problem (cf. 2.1) and the Suliński-Anderson-Divinsky Problem (cf. Theorem 1.1.1). In fact all his papers on radicals contained solutions of open problems or substantial new results. Kostia's reputation was indisputable; talking once on a surprising new result, someone asked who had proved it. "Beidar, who else?"—replied Arthur Sands. And this was typical, indeed, speaking to colleagues and to his collaborators, everybody agrees that Kostia was the guy who could answer all problems affirmatively or in the negative.

It is the purpose of this paper to give an account on a segment of Kostia Beidar's contribution to algebra, namely to radical theory and related topics.

The main objective of this survey is radical theory, therefore, in the sequel we recall some fundamental definitions of radical theory.

In what follows $I \lhd A$ ($I \lhd_l A$, $I \lhd_r A$) denotes that I is an ideal (left ideal, right ideal, respectively) of a ring A.

A class γ of rings is called a *radical class* or shortly a *radical* in the sense of Kurosh and Amitsur, if

  i) γ is closed under homomorphisms,
 ii) $\gamma(A) = \sum(I \lhd A \mid I \in \gamma) \in \gamma$ for all rings A,
iii) $\gamma(A/\gamma(A)) = 0$ for all rings A.

The *semisimple class* $\mathcal{S}\gamma$ of a radical γ is defined as

$$\mathcal{S}\gamma = \{A \mid \gamma(A) = 0\}.$$

Rings belonging to γ are called γ-radical and those belonging to $\mathcal{S}\gamma$, γ-semisimple.

The *upper radical* $\mathcal{U}\varrho$ of (or, determined by) a class ϱ of rings is defined as the largest radical class γ such that all rings from ϱ are γ-semisimple. It is well known that in the class of associative rings $\mathcal{U}\varrho$ always exists and in many cases (not only for associative rings), for instance, if ϱ is a regular class,

$$\mathcal{U}\varrho = \{A \mid A \text{ has no nonzero homomorphic image in } \varrho\}.$$

A class ϱ of rings is called a *regular class* if $0 \neq I \lhd A$ and $A \in \varrho$ then I has a nonzero homomorphic image in ϱ. A class ϱ of rings is said to be *hereditary*, if $I \lhd A \in \varrho$ implies $I \in \varrho$. A hereditary class is always regular.

A class γ of rings is said to be *left hereditary* if $L \lhd_l A \in \gamma$ implies $L \in \gamma$. Right hereditary classes are defined correspondingly.

A radical γ is said to be *left strong*, if $L \lhd_l A$ and $L \in \gamma$ imply $L \subseteq \gamma(A)$. Right strong radicals are defined similarly. We say that γ is *strict* if $\gamma(A)$ contains every subring S of A with $S \in \gamma$. We say that γ is *left (right) stable* if its semisimple class $\mathcal{S}\gamma$ is left (right) hereditary.

Throughout the paper, except the last section, all considered rings are associative (though sometime we make short comments concerning other situations) but not necessarily with unity element. The ring obtained from a given ring A by adjoining a unity element is denoted by A^1. The ring of integers is denoted by $\mathbb{Z}$.

1. General radical theory

1.1. The Suliński-Anderson-Divinsky problem. The smallest radical class containing a given class δ of rings, called the *lower radical* of (or, determined by) δ and denoted $\mathcal{L}\delta$, can be constructed as follows.

$$\delta_1 = \{A \mid A \text{ is a homomorphic image of a ring in } \delta\}.$$

Assuming δ_μ has been defined for every ordinal $\mu < \lambda$, we define

$$\delta_\lambda = \left\{ A \;\middle|\; \begin{array}{l} \text{every nonzero homomorphic image of } A \text{ has} \\ \text{a nonzero ideal in } \delta_\mu \text{ for some } \mu < \lambda \end{array} \right\}.$$

Then the lower radical $\mathcal{L}\delta$ is the union of all the classes δ_λ.

This construction is valid for any universal class of rings (not necessarily associative). Suliński, Anderson and Divinsky [120] proved that the lower radical construction for associative rings terminates at the first limit ordinal ω, that is, $\mathcal{L}\delta = \delta_\omega$. Further, they proved that for a *hereditary class* δ which contains all zero-rings, $\mathcal{L}\delta = \delta_2$. Hoffman and Leavitt [70] proved that if the class δ consists of idempotent rings, then $\mathcal{L}\delta = \delta_2$.

Suliński, Anderson and Divinsky [120] posed the following problem in 1966:

> *Are there examples of classes of rings for which the lower radical*
> *construction terminates in precisely* 3, 4, ..., ω *steps?*

Armendariz and Leavitt [11] and Watters [129] showed that for a hereditary and homomorphically closed class δ the lower radical construction terminates at the third step, $\mathcal{L}\delta = \delta_3$, and gave example of such a class δ for which $\mathcal{L}\delta \neq \delta_2$. Stewart [119] proved that if δ is a homomorphically closed class of nilpotent rings, then $\mathcal{L}\delta = \delta_3$ and this result is sharp. Heinicke [68] constructed a class δ of rings for which $\mathcal{L}\delta = \delta_\omega \neq \delta_n$ for all finite n.

It was Beidar [18] who answered the Suliński-Anderson-Divinsky problem affirmatively in 1982. He explored some hidden properties of Gaussian integers (a ring so well known that there is hardly anything new that can be said on it) by examining the accessible subring structure of the Gaussian integers.

Let $\mathbb{Q}(i)$ be the extension of the rationals by $i = \sqrt{-1}$, $G = \{a + bi \mid a, b \in \mathbb{Z}\}$ the ring of Gaussian integers, p a prime of the form $p = 4k + 3$. Further, define

$$A_0 = G \quad \text{and} \quad A_n = \{pa + p^n bi \mid a, b \in \mathbb{Z}\}$$

for $n = 1, 2, \ldots$.

i) If $0 \neq I \triangleleft A_n$ and $f : A_n \to \mathbb{Q}(i)$ is a ring homomorphism, then A_n/I is finite and $f(A_n) = 0$ or A_n.

ii) If $A_{n+m} \triangleleft A_n$, then either $m = 0$ or $m = 1$.

iii) If R is a subring of $\mathbb{Q}(i)$ such that $A_n \triangleleft R$, then $R = A_{n+1}$ or $R = A_n$ or $1 \in R$.

iv) Let

$$A_n \cong B_m \triangleleft B_{m-1} \triangleleft \cdots \triangleleft B_1 \triangleleft B_0 \subseteq \mathbb{Q}(i)$$

be a finite chain without repetition. If $1 \notin B_0$, then

$$A_n = B_m, \quad A_{n-1} = B_{m-1}, \ldots, A_{n-m+1} = B_1, \quad A_{n-m} = B_0,$$

and if $1 \in B_0$, then

$$A_n = B_m, \quad A_{n-1} = B_{m-1}, \ldots, A_{n-m+1} = B_1.$$

Using these properties, Beidar [18] proved the following

Theorem 1.1.1. *Let $n > 0$ be a fixed integer or ω and*

$$\delta = \{A_n\} \cup \{\text{all rings with torsion additive group}\}.$$

Then $\mathcal{L}\delta = \delta_{n+1} \neq \delta_n$.

Refining the method used in proving Theorem 1.1.1, in [19] Beidar *sharpened that result to any variety of alternative rings containing the ring of integers.*

Guo Jin-yun [66] solved the Suliński-Anderson-Divinsky problem in 1987 by using the ring

$$A_n = \left\{ \sum_{i=1}^{n} a_i x^{in} + \sum_{j>0} b_j x^{n^2} + j \;\middle|\; a_i, b_j \in F_p \right\}$$

of polynomials over the prime field F_p.

Both Beidar and Guo settled the Suliński-Anderson-Divinsky problem within the variety of *commutative*, associative rings.

Andruszkiewicz and Puczyłowski [9] settled the Suliński-Anderson-Divinsky problem in a more general setting which included also the case of the left strong lower radical construction. In [10] they have shown that starting from any commutative noetherian integrally closed domain with unity, classes δ can be constructed such that the lower radical construction terminates in exactly n or ω steps.

Studying the accessible subring structure Andruszkiewicz [7] got the following result: *the positive solution of the Suliński-Anderson-Divinsky problem is equivalent to the question whether for every $n \geq 3$ there exists a homomorphically closed class δ of rings such that if a ring A has a nonzero n-accessible subring in δ, then A has a nonzero $(n-1)$-accessible subring in δ.* Andruszkiewicz [8] studied also the termination of the essential cover construction.

The Suliński-Anderson-Divinsky problem has a positive solution also for algebras over a commutative ring with unity (L'vov and Sidorov [85], and Watters [130]) and for semigroups (Sidorov [115]).

1.2. Special radicals. The original aim of introducing radicals was to decompose the radical-free, that is, semisimple rings into a (sub)direct sum of "good" rings.

A subring I of a ring A is *essential* in A, if $I \cap K \neq 0$ for every nonzero ideal K of A. To denote that an ideal I a ring A is essential in A, we write $I \triangleleft \cdot A$. A class ϱ of rings is said to be *closed under essential extensions*, if $I \triangleleft \cdot A$ and $I \in \varrho$ imply $A \in \varrho$. A hereditary class ϱ of prime rings which is closed under essential extensions is called a *special class*.

The upper radical $\gamma = \mathcal{U}\varrho$ of a special class ϱ is called a *special radical*. Andrunakievich [3] proved that *every special radical $\gamma = \mathcal{U}\varrho$ is hereditary and every γ-semisimple ring A is a subdirect sum of rings taken from the special class ϱ.* Thus, special radicals are the "good" ones; examples are the Baer (prime), locally nilpotent, nil, Jacobson and Brown-McCoy radicals.

We say that a radical γ has the *intersection property relative to the class* ϱ, if

$$\gamma(A) = \cap(I \lhd A \mid A/I \in \varrho) \quad \text{for all rings } A.$$

Every special radical $\gamma = \mathcal{U}\varrho$ has the intersection property relative to the special class ϱ. Beidar [22] characterized special radicals as follows.

Theorem 1.2.1. *A radical γ is special if and only if it is the upper radical $\gamma = \mathcal{U}\varrho$ determined by some class ϱ of prime rings, and has the intersection property relative to the class ϱ.*

Theorem 1.2.1 is a deep result with important consequences.

If a radical $\gamma = \mathcal{U}\varrho$ has the intersection property relative to a class ϱ consisting of simple prime rings, then by Theorem 1.2.1, γ is a special radical. Then by a result of Andrunakievich, the rings in ϱ have a unity element. Thus, the upper radical $\gamma = \mathcal{U}\varrho$ of a class ϱ of simple prime rings has the intersection property relative to ϱ if and only if every ring in ϱ has a unity element. This is a theorem of Leavitt [79], who proved it in a much more complicated way (nevertheless, with his method Leavitt solved simultaneously also another difficult problem: *there exists a class ϱ of simple prime rings not all with unity element such that the upper radical $\mathcal{U}\varrho$ is hereditary*).

For every radical class γ Ortiz [90] considered a radical class γ_x defined in terms of polynomials $A[x]$ which is in Gardner's [59] formulation

$$\gamma_x = \{A \mid A[x] \in \gamma\}.$$

Ortiz [90] proved that γ_x is a special radical whenever so is γ. The proof can be substantially simplified by using Beidar's characterization of special radicals (see [65]).

Applying appropriate constructions and the powerful method of generalized identities in ring theory, Arnautov, Beidar, Glavatskiĭ and Mikhalev [14] extended Theorem 1.2.1 to associative Hausdorff R-algebras over a commutative ring R with discrete topology (see also [15]).

1.3. Disjoint special classes defining the same radical. Given a special class ϱ, the special radical $\gamma = \mathcal{U}\varrho$ can be obtained also as $\gamma = \mathcal{U}(\mathcal{S}\gamma \cap \mathcal{P})$, where $\mathcal{S}\gamma \cap \mathcal{P}$ is the special class of all γ-semisimple prime rings. It may happen that ϱ is properly contained in $\mathcal{S}\gamma \cap \mathcal{P}$. It may also happen that a special class ϱ does not contain simple rings and in this case the semisimple class $\mathcal{S}\gamma$ of the special radical $\gamma = \mathcal{U}\varrho$ does not contain subdirectly irreducible rings (see Leavitt and Watters [80]). Now, every ring $A \in \mathcal{S}\gamma$ is a subdirect sum of rings taken from ϱ as well as a subdirect sum of subdirectly irreducible rings, none in $S\gamma$. Clearly, these two subdirect decompositions are totally different. In that context Gardner [62] posed the question: *do there exist disjoint special classes ϱ_1 and ϱ_2 of rings such that $\mathcal{U}\varrho_1 = \mathcal{U}\varrho_2$?*

Beidar [23] gave an affirmative answer to this question.

Theorem 1.3.1. *Let C be the algebraic closure of the field of rational numbers, $A = C[x_1, x_2, \dots]$ the ring of polynomials over C in commuting indeterminates and I the ideal of A generated by the polynomial $x_1^2 + x_2^2 - 1$. If $\varrho(A)$ and $\varrho(A/I)$ are the special classes generated by A and A/I, respectively, then*

(i) $\mathcal{U}\varrho(A) = \mathcal{U}\varrho(A/I)$,
(ii) $\varrho(A) \cap \varrho(A/I) = \{0\}$.

1.4. Non-special supernilpotent radicals. A radical class γ is said to be *supernilpotent*, if γ is hereditary and contains all nilpotent rings. Obviously, every special radical is supernilpotent. A hereditary class ϱ of semiprime rings which is closed under essential extensions, is called a *weakly special class*. Ryabukhin [106] proved that a radical γ is supernilpotent if and only if γ is the upper radical $\gamma = \mathcal{U}\varrho$ of a weakly special class ϱ. Is every supernilpotent radical a special one, or do there exist non-special supernilpotent radicals? Ryabukhin [106] considered the class ϱ of all Boolean rings A having only infinite nonzero ideals, and showed that ϱ is a weakly special class, but $\mathcal{U}\varrho$ is not a special radical. Van Leeuwen and Jenkins [81] constructed infinitely many non-special supernilpotent radicals such that their semisimple classes contain all fields. Ryabukhin [107] showed that every supernilpotent radical different from the Baer radical is the union of non-special supernilpotent radicals.

F. A. Szász [122] posed the problem: *are there infinitely many non-special supernilpotent radical classes $\gamma_1, \gamma_2, \dots$ such that $S\gamma_m \cap S\gamma_n = \{0\}$ for $m \neq n$?* An affirmative answer was given by Beidar and Salavová [41]. For every n they considered the classes ν_n of all semiprime rings A such that $I \lhd A \in \nu_n$ implies that I is not a prime ring and $d(I) = n$, where

$$d(I) = \min \left\{ k \in \mathbb{N} \, \middle| \, \begin{array}{l} \text{the standard polynomial identity of} \\ \text{degree } 2k \text{ is an identity of } I \end{array} \right\}.$$

Theorem 1.4.1 ([41]). *Each class ν_n is a weakly special class such that $\gamma_n = \mathcal{U}\nu_n$ is a non-special supernilpotent radical and $S\gamma_m \cap S\gamma_n = \{0\}$ for $m \neq n$.*

More examples of non-special supernilpotent radicals with some additional requirements were given by Gardner and Stewart [63], France-Jackson [58] and Tumurbat [125]. It was shown in [48] that *the Kasch radical $\mathcal{K}$ is a non-special supernilpotent radical* (see Theorem 2.4.3).

1.5. Dependence and independence among radicals involving one-sided ideals.
Many of Kostia's studies concerned radicals and one-sided ideals of associative rings. In particular he studied relations among left and right strong radicals and left and right hereditary radicals.

In [56] Divinsky, Krempa and Suliński (see also Sands [110]) proved that *if C is a regular class of rings then the upper radical $\mathcal{UC}$ determined by C is left strong if and only if every nonzero left ideal of an arbitrary ring from C has a nonzero homomorphic image in C.*

Denote for a given subset X of a ring A the left and right annihilators of X in A by $l_A(X)$ and $r_A(X)$, respectively.

In [42] Beidar and Salavova proved that following theorem.

Theorem 1.5.1. *If a class $\mathcal{P}$ of semiprime rings satisfies*

1. *If $0 \neq L \lhd_l A$ and $A \in \mathcal{P}$, then $L/r_L(L) \in \mathcal{P}$;*
2. *If $I \lhd A$, $l_A(I) = 0$, A is a semiprime ring and $I \in \mathcal{P}$, then $A \in \mathcal{P}$,*

then the upper radical $\mathcal{UP}$ of $\mathcal{P}$ is left strong and supernilpotent.

Conversely, if a radical is left strong and supernilpotent, then its semisimple class satisfies 1. *and* 2.

They also obtained a dual result.

Theorem 1.5.2 ([42]). *If a class $\mathcal{P}$ of semiprime rings satisfies*

1'. *If $0 \neq L \lhd_l A$, A is semiprime, $l_R(L) = 0$ and $L/r_L(L) \in \mathcal{P}$, then $A \in \mathcal{P}$;*
2'. *$\mathcal{P}$ is hereditary,*

then the upper radical $\mathcal{UP}$ is left hereditary and supernilpotent.

Conversely, if a radical is left hereditary and supernilpotent, then its semisimple class satisfies 1'. *and* 2'.

A left strong and left hereditary radical containing the prime radical is called N-*radical* [108].

Combining Theorems 1.5.1 and 1.5.2 one obtains a characterization of upper radicals which are N-radicals.

In [111] Sands proved that *every left or right strong and left or right hereditary radical has all these properties.* He asked whether one can extend this result by proving that every left strong radical is right strong and/or every left hereditary radical is right hereditary. Applying Theorems 1.5.1 and 1.5.2, Beidar gave in [19] negative answers to both these questions. He proved there the following theorem.

Theorem 1.5.3. *Let D be a simple domain with unity element which is not right noetherian and let $\mathcal{H}$ be the class of subdirectly irreducible rings whose hearts are isomorphic (as rings) to some left ideals of D. Then the upper radical $\mathcal{UH}$ is left but not right strong. The upper radical u^* determined by the class of prime subdirectly irreducible rings which are not $\mathcal{UH}$-semisimple is left but not right hereditary.*

As it was observed later left but not right strong radicals can be found among some classical radicals.

A ring A is called [69] *left strongly prime* if every nonzero ideal of A contains a finite set F such that $l_A(F) = 0$. Right strongly prime rings are defined similarly. The *left (right) strongly prime* radical is defined as the upper radical of the class of left (right) strongly prime rings.

In [95] it was proved that the left strongly prime radical is left but not right strong.

A radical γ is called *principally left hereditary* if for every element a of an arbitrary γ-radical ring A, $Aa \in \gamma$. Principally right hereditary radicals are defined dually.

Sands [109] proved that every left strong and principally left hereditary radical is right strong and principally right hereditary. He asked (see also [112])

  (a) *If a radical is principally right hereditary and left strong, is it then principally left hereditary?*

Jaegermann and Sands [71] proved that a radical is left stable and principally right hereditary if and only if it is right stable and principally right hereditary. They asked

  (b) *If a radical is principally right hereditary and left stable, is it then principally left hereditary?*

Sands [112] asked

  (c) *If a radical is principally right hereditary, is it then principally left hereditary?*

In [25] Beidar answered all these questions. In fact he got much more. The following are his main results obtained in [25] concerning that matter.

A radical γ is called [25] *left almost hereditary* if for arbitrary $L \triangleleft_l A$, $\gamma(A)L \in \gamma$ and it is called *left almost stable* if for arbitrary $L \triangleleft_l A$, we have $L\gamma(L) \subseteq \gamma(A)$.

Theorem 1.5.4 ([25]). *For a given radical γ the following conditions are equivalent*

1. γ *is left or right strong and principally left or right hereditary;*
2. γ *is left and right strong and principally left and right hereditary;*
3. γ *is left or right almost stable and left or right almost hereditary;*
4. γ *is left and right almost stable and left and right almost hereditary.*

This in particular gives a positive answers to (a).

Theorem 1.5.5 ([25]). *For a given radical γ the following conditions are equivalent*

1. γ *is left or right stable and principally left or right hereditary;*
2. γ *is left and right stable and principally left and right hereditary;*
3. γ *is left or right stable and left or right almost hereditary;*
4. γ *is left and right stable and left and right almost hereditary.*

This in particular gives a positive answer to (b).

Theorem 1.5.6 ([25]). *Every left almost hereditary radical is principally left hereditary. Every left almost stable radical is left strong.*

Theorem 1.5.7 ([25]). *There exists a right almost hereditary radical consisting of idempotent rings, which is not principally left hereditary.*

Clearly Theorems 1.5.6 and 1.5.7 give a negative answer to (c).

In [93] it was proved that a radical containing the prime radical is principally left hereditary if and only if it is principally right hereditary as well as that every left hereditary radical is principally right hereditary. Extending the method applied in [93] Sands proved in [112] that a principally left hereditary radical, which is hereditary, is also left principally hereditary. He asked: *does there exist a hereditary and principally left hereditary radical which is not left hereditary?* An example of such a radical was constructed by Beidar in [23]. Answering another question of Sands, he proved there also that *every left hereditary radical contained in the class of idempotent rings is right hereditary.*

It is clear that a radical γ is left strong if and only if no γ-semisimple ring contains nonzero left γ-ideals. In [103] Puczyłowski and Zand started studies of radicals satisfying the counterpart of this condition with the terms "semisimple"

and "radical" interchanged. These radicals were called in [103] *left subhereditary.*
More precisely, a radical γ is left subhereditary if and only if for every nonzero left
ideal L of an arbitrary γ-radical ring A, $\gamma(L) \neq 0$. Right subhereditary radicals
are defined similarly. Some questions concerning such radicals were considered
much earlier by Sands. For instance in [62] he raised a problem (still open) which
can be formulated in terms of subhereditary radicals as follows: *is every left subhereditary and right hereditary radical left hereditary?* Sands also asked: *is the
Brown-McCoy radical subhereditary?*

Left (right) subhereditary radicals are dual to left (right) strong radicals in a
similar way as left (right) hereditary radicals are dual to left (right) stable radicals.
As we already mentioned, Sands [111] proved that if a radical is left or right strong
and left or right hereditary then it satisfies all these conditions. In [103] it was
asked whether the dual result holds as well, that is, *whether if a radical is left or
right subhereditary and left or right stable then it has all these properties.*

In [103] partial results concerning the above mentioned problems were obtained
and new questions were asked. Some of these questions were answered by Beidar,
Ke and Puczyłowski in [34]. In particular it was shown there that

a) *the Brown-McCoy radical is neither left nor right subhereditary (that is, a
ring possessing a Brown-McCoy semisimple essential left or right ideal need
not be Brown-McCoy semisimple; but*, as pointed out in [126], *it is Behrens
semisimple);*
b) *the left strongly prime radical is not left subhereditary;*
c) *every radical which is left or right subhereditary and left or right stable satisfies all these conditions.*

Below we quote two most substantial results from [34], which in particular
easily give a)–c).

For a given ring A and a non-empty set X denote by $M_X^c(A)$ (respectively,
$M_X^r(A)$) the ring of matrices over A indexed by elements from X, which have
only finitely many nonzero columns (respectively, rows).

For a given set X denote by $\mathbb{Z}\langle X \rangle$ (respectively, $\mathbb{Z}^*\langle X \rangle$) the free ring without
unity (respectively, the free ring with unity) in indeterminates from the set X.

Theorem 1.5.8 ([34]). *For every non-empty set X, $\mathbb{Z}\langle X \rangle$ is isomorphic to an
essential right ideal of $M_X^c(\mathbb{Z}\langle X \rangle)$ and to an essential left ideal of $M_X^r(\mathbb{Z}\langle X \rangle)$.*

Theorem 1.5.9 ([34]). *For a given radical γ the following conditions are equivalent*

(i) γ is right subhereditary and for every infinite set X the ring $M_X^c(\mathbb{Z}\langle X\rangle)$ is γ-radical;

(ii) γ is left subhereditary and for every infinite set X the ring $M_X^r(\mathbb{Z}\langle X\rangle)$ is γ-radical;

(iii) γ is hereditary and all γ-semisimple rings satisfy a common polynomial identity.

However these results were not strong enough to determine whether the left strongly prime radical is right subhereditary. This question was answered by Beidar, Ke and Puczyłowski [35] in the negative. It was a consequence of the following extensions of Theorem 1.5.9.

Theorem 1.5.10 ([35]). *The conditions of Theorem 1.5.9 are equivalent to each of the following:*

(a) γ is left subhereditary and, for arbitrary infinite sets X and Y, the ring $M_X^r(\mathbb{Z}\langle Y\rangle)$ is γ-radical;

(b) γ is left subhereditary and, for every countable set X and every infinite set Y, the ring $M_X^r(\mathbb{Z}\langle Y\rangle)$ is γ-radical;

(a') γ is right subhereditary and, for arbitrary infinite sets X and Y, the ring $M_X^c(\mathbb{Z}\langle Y\rangle)$ is γ-radical;

(b') γ is right subhereditary and, for every countable set X and every infinite set Y, the ring $M_X^c(\mathbb{Z}\langle Y\rangle)$ is γ-radical.

Theorem 1.5.11 ([35]). *The conditions in Theorem 1.5.9 are equivalent to each of the following:*

(a) γ is left subhereditary and, for arbitrary infinite sets X and Y, the ring $M_X^c(\mathbb{Z}\langle Y\rangle)$ is γ-radical;

(b) γ left subhereditary and, for every infinite set X, the ring $M_X^c(\mathbb{Z}\langle X\rangle)$ is γ-radical;

(a') γ is right subhereditary and, for arbitrary infinite sets X and Y, the ring $M_X^r(\mathbb{Z}\langle Y\rangle)$ is γ-radical;

(b') γ is right subhereditary and, for every infinite set X, the ring $M_X^r(\mathbb{Z}\langle X\rangle)$ is γ-radical.

A substantial role in obtaining these results played the following theorem which might be useful in other studies of rings.

Let S be a nonempty set and let Y be any set disjoint to S. Denote by A_S the left ideal of $M_S^c(\mathbb{Z}^*\langle S \cup Y\rangle)$ consisting of all matrices whose entries in the s-column belong to $\mathbb{Z}\langle S \cup Y\rangle s$, $s \in S$ and set $I_S = \sum_{s\in S} s\mathbb{Z}^*\langle S \cup Y\rangle$.

Theorem 1.5.12 ([35]). *Let X be an infinite set and Y any set disjoint to X. The ring A_X is a subdirect sum of rings A_S, where S runs over all finite subsets of X, and each A_S contains an essential left ideal isomorphic (as a ring) to I_S.*

1.6. Explicit description of radical classes with semisimple essential cover.
Classes of rings which are simultaneously radical and semisimple classes were explicitly determined by Stewart [118]: *every radical and semisimple class ϱ ($\neq \{all\ rings\}$) is the class of all subdirect sums of rings taken from a finite family of finite fields closed under subfields.* It is easily seen that these classes are subvarieties of the variety of associative rings. The defining identities of these varieties were determined by Gardner and Stewart [64]. Loi [82] proved that *the radical and semisimple classes are exactly the radical classes closed under essential extensions.*

Birkenmeier [51] studied classes σ of rings which are essential extensions of supernilpotent radicals ϱ:

$$\sigma = \mathcal{E}\varrho = \{A \mid \exists\, I \lhd \cdot A \text{ with } I \in \varrho\},$$

that is, σ is the *essential cover* $\mathcal{E}\varrho$ of ϱ.

Let us observe that the essential cover of a class of rings is not necessarily closed under essential extensions, and the essential cover $\mathcal{E}\varrho$ properly contains the supernilpotent radical class ϱ in view of Loi [82], but $\varrho = \mathcal{E}\varrho$ for every radical and semisimple class ϱ.

Birkenmeier [51] found that the essential cover $\mathcal{E}\varrho$ of a supernilpotent radical is *nearly* a semisimple class: it is hereditary, closed under extensions, finite subdirect sums and arbitrary direct sums and direct products. He raised the question: *which radical classes ϱ have semisimple essential covers $\mathcal{E}\varrho$?* This question was addressed in Birkenmeier [52] and Wu [132]. A complete solution of Birkenmeier's problem was given by Beidar, Fong, Ke and Shum [27] (see also Beidar, Fong and Ke [30]). In formulating this result, we say that a semiprime ring A is an *F-ring* with respect to a simple prime ring F, if each of its prime homomorphic images is isomorphic to F.

Theorem 1.6.1. *For a radical γ the following conditions are equivalent:*

(i) *the essential cover $\mathcal{E}\gamma$ is a semisimple class;*

(ii) *either γ is the upper radical generated by a finite number of finite simple prime rings, or there exist finitely many matrix rings $M_{n_i}(F_i)$, $i = 1, \ldots, k$ over finite fields F_i such that γ is the lower radical generated by all rings $M_{n_i}(R_i)$, $i = 1, \ldots, k$, where each R_i is an F_i-ring;*

(iii) *either* γ *or* $\mathcal{UE}\gamma$ *is the upper radical of a finite set of finite simple rings with unity element.*

From (ii) and (iii) one sees that radical classes having semisimple essential covers are supernilpotent or subidempotent.

Birkenmeier and Wiegandt [53] gave another characterization of radicals with semisimple essential cover. *The essential cover of a radical class* γ *is a semisimple class if and only if* γ *is hereditary and has a complement in the lattice of all hereditary radicals.* The complemented supernilpotent radicals were already described by Snider [116] as in Theorem 1.6.1.

Next, we mention also another difficult result of Kostia.

F. A. Szász [122] posed the problem to characterize the *additive radicals* γ, that is, $\gamma(I + J) = \gamma(I) + \gamma(J)$ for arbitrary ideals I and J of an arbitrary ring A. Radicals γ with homomorphically closed semisimple class $\mathcal{S}\gamma$ (that is, $\mathcal{S}\gamma$ is a radical semisimple class) are examples for additive radicals. Mobilizing quite a heavy machinery Beidar and Salavová [43] answered this problem. They obtained the following result.

Theorem 1.6.2. *The following conditions are equivalent:*

(i) γ *is an additive radical;*
(ii) *either there exists an* $n > 1$ *such that the semisimple class* $\mathcal{S}\gamma$ *satisfies the identity* $x^n - x = 0$, *or* γ *consists of idempotent rings (that is,* γ *is hypoidempotent) and* $I + J = A \in \gamma$ *implies* $\gamma(I) + \gamma(J) = A$.

For a hereditary radical γ *the following are equivalent:*

(i) γ *is an additive radical;*
(ii) *either there exists an* $n > 1$ *such that* $\mathcal{S}\gamma$ *satisfies the identity* $x^n - x = 0$, *or* γ *is a hypoidempotent radical.*

1.7. Lattices of radicals. It is well known that the class of all radicals of associative rings forms a complete lattice with respect to inclusion. The same concerns radicals of some specific types, e.g., hereditary radicals. Systematic studies of lattices of radicals of associative rings were started by Snider in [117]. His main results concerned the lattice of hereditary radicals. In particular he showed that a hereditary radical is an atom of this lattice if and only if it is the lower radical determined by a simple ring (not necessarily idempotent). It is not hard to notice that the lower radical determined by a simple ring with unity is an atom of the lattice of all radical. In [6] Andrunakievich and Ryabukhin asked for a description of all atoms in the lattice of all radicals as well as for a description of simple rings without unity element which determine atoms. At that time no such a simple ring was known. In [61] Gardner observed that if a simple ring P satisfies the property

$$\text{if } P \triangleleft A \text{ and } A/P \cong P, \text{ then } A \cong P \oplus P,$$

then the lower radical determined by P is an atom of the lattice of all radicals. An example of a simple ring which does not determine an atom was given by Gardner in [60]. Examples of simple rings without unity element determining atoms were given independently by Gardner [61], Korolczuk [75] and Beidar [19]. The first two examples were based on [79]. Beidar proved the following result.

Theorem 1.7.1 ([19]). *If D is a simple domain in which left ideals are linearly ordered by inclusion, then the lower radical determined by D is an atom in the lattice of all radicals.*

An example of a ring without unity element satisfying the assumptions of Theorem 1.7.1 was constructed by Dubrovin in [57].

In [61] Gardner proved that *in the class of all, not necessarily associative, rings the lower radical determined by an arbitrary simple ring is not an atom in the lattice of all radicals.*

A significant contribution to the problem of a description of atoms of the lattice of all radicals was made by Beidar in [20].

Let L be a left ideal of a ring A and let $f_L : L \to \text{End}(L_L)$ be the ring homomorphism defined by $f_L(a)b = ab$ for $a \in A$, $b \in L$. The subring

$$MI(L) = \{\varphi \in \text{End}(L_L) \mid f_L(L)\varphi \subseteq f_L(L)\}$$

of $\text{End}(L_L)$ is called *the maximal left idealizer of L.*

Theorem 1.7.2 (Beidar [20]). *If A is a nonzero idempotent ring such that for every nonzero homomorphic image Q of A*

  (i) *there is no nonzero ring homomorphism $A \to MI(Q)/f_Q(Q)$,*

  (ii) *A belongs to the lower radical determined by Q,*

then the lower radical determined by A is an atom of the lattice of all radicals.

Applying this result Beidar proved that *if S is a simple nonartinian ring with unity which is not a domain, then every maximal left ideal of S is a simple ring which determines an atom.* A similar result was independently obtained by different methods in [97].

Let F be a finite prime field and let H be a dense subgroup of the additive group of real numbers. Denote by $F(H)$ the F-algebra with basis

$$\{x_h \mid h \in H \mid 0 < h < 1\}$$

and multiplication

$$x_h x_g = \begin{cases} x_{h+g} & \text{if } h + g < 1, \\ 0 & \text{otherwise.} \end{cases}$$

Applying Theorem 1.7.2 Beidar [20] showed that *the lower radical determined by $F(H)$ is an atom or it contains no atoms, the former being the case if and only if there is a positive integer c with $cH = H$*. If, for example, H is the subgroup generated by the inverses of the primes, then the lower radical determined by $F(H)$ contains no atoms. These studies were continued in [101].

A radical γ is called *complemented* if there exists a radical γ', called a *complement* of γ, such that $\gamma \cap \gamma' = 0$ and the lower radical determined by $\gamma \cup \gamma'$ is equal to the class of all rings, i.e., γ is complemented as an element of the lattice of all radicals.

Complemented elements of the lattice of hereditary radicals were described by Snider in [116]. The problem of finding a description of complemented radicals in the lattice of all radicals existed in the folklore for many years. It was solved by Beidar, Fong and Ke in [30]. This very deep result implies in particular that a radical γ is complemented if and only if γ or its complement is the upper radical determined by a finite family of matrix rings over finite fields. This paper contains also many other results (see for instance Theorem 1.6.1) which were applied in further studies. Among them the following one is of particular importance.

Theorem 1.7.3 ([30]). *Let C be an abstract (i.e., closed under isomorphisms) class of ring. If C is either hereditary or homomorphically closed, then either rings from C satisfy a common proper polynomial identity or there exists an integer $m \neq \pm 1$ such that the free $\mathbb{Z}_m$-algebra $\mathbb{Z}_m\langle X\rangle$ in an infinite set X of generators is a subdirect sum of rings from C, where $\mathbb{Z}_m$ is the ring of integers mod m.*

The class of hereditary radicals is not only a lattice with respect to inclusion but is also a sublattice of the lattice of all radicals. Since [91] *the intersection of any family of left (right) strong radicals is a left (right) strong radical*, the class of left (right) strong radicals is a lattice with respect to inclusion. However, this is not a sublattice of the lattice of all radicals (cf. [91]). In [93] it was proved that the class of left (right) strong and hereditary radicals is already such a sublattice. In that context it was natural to ask whether some related classes of radicals are sublattices of the lattice of all radicals. For a long time it was unknown whether it is so for the class of left and right strong radicals [96, 99, 111, 112] and for the class of left strong radicals containing the prime radical [96, 99]. In [28] Beidar, Fong and Wang constructed an example giving negative answers to both of these questions. The idea of the example is the following.

Let R be the ring of 2×2-matrices over the ring $\mathbb{Z}$ of integers, $A = \mathbb{Z}e_{12} + 2R$ and F be a two-element field, where e_{12} is the usual matrix unit. Now, if α is the smallest left and right strong radical containing the prime radical and A, and γ is the smallest left and right strong radical containing the prime radical and F, then the lower radical determined by $\alpha \cup \gamma$ is not right strong.

Moreover Beidar, Fong and Wang showed that R contains a right ideal J, which is a semiprime ring such that $\alpha(J) \neq 0$. This gave a negative answer to a question posed in [96].

In [103] it was shown that the class of left (right) subhereditary radicals is a sublattice of the lattice of all radicals. However, it was unknown whether this lattice is complete. A negative answer to this question was given by Beidar, Ke and Puczyłowski in [35] as a consequence of the following results proved there.

Theorem 1.7.4. *Let $\mathcal{M}$ be a class of reduced rings. The upper radical $\mathcal{UM}$ determined by the class $\mathcal{M}$ is left (right) subhereditary if and only if all rings in $\mathcal{M}$ satisfy a common polynomial identity.*

In [103] it was shown that besides lower radicals determined by simple rings with zero multiplication all the other atoms in the lattice of left subhereditary radicals are uniquely determined by some specific simple domains. It is an open problem (cf. Question 1 in [103] and Problem 4.4 in [34]) whether all these domains are division rings. A related, interesting on its own, problem (cf. Question 2 in [103] and Question 5 in [34]) is: *are there simple domains not being division rings, which are isomorphic (as rings) to all their idempotent left ideals?* A partial answer to this question gives the following result.

Theorem 1.7.5 (Beidar, Ke and Puczyłowski [35]). *Suppose that A is a simple domain with unity and $0 \neq L \lhd_l A$. If L is isomorphic (as a ring) to each of its nonzero idempotent left ideals, then A is a division ring.*

The proof of this theorem is based on some very interesting results obtained by Beidar in [19]. Let us quote two of them which might be useful in other studies of rings.

Theorem 1.7.6 ([19]). *If a nonzero right ideal M of a simple ring A with unity is ring-isomorphic to a left ideal of A, then M is a finitely generated projective right ideal of A.*

Theorem 1.7.7 ([19]). *Suppose that A and B are simple rings with unity having ring-isomorphic left ideals L and M, respectively. Then the isomorphism of L to M extends to an isomorphism of A to B.*

2. Concrete radicals and structure theory

2.1. On the Jacobson radical of finitely generated algebras. One of first famous results obtained by Kostia concerned the Jacobson radical of finitely generated algebras over countable fields.

In [1] Amitsur proved that if A is an algebra over a field F and $\dim_F A <$ card$F$, then the Jacobson radical of A is nil. This result applies when F is uncountable and A is a finitely generated F-algebra and shows that Koethe's [76] problem has a positive solution in the class of algebras over arbitrary uncountable field. Amitsur asked whether this result can be extended to all fields (which would give a positive solution of the Koethe's problem in general). This question was for a long time open. In [17] Beidar showed that the answer is negative. He employed the following general result due to Markov.

Theorem 2.1.1 ([86]). *Suppose that $\mathcal{R}$ is an abstract class of algebras over a field F such that*

   (i) $\mathcal{R}$ *contains an algebra A with* $\dim_F A \leq \aleph_0$;
   (ii) *If* $I \triangleleft R,\ I^2 = 0$ *and* $R/I \in \mathcal{R}$ *then* $R \in \mathcal{R}$.

Then there exists an F-algebra with unity generated by two elements containing a nonzero left ideal belonging to $\mathcal{R}$.

In [17] Beidar made the following observation. Let F be a field with card$F \leq \aleph_0$ and A be the Jacobson radical of the localization of the polynomial ring $F[x]$ at a maximal ideal of $F[x]$. Clearly A is a domain and $\dim_F A \leq \aleph_0$. Hence the class $\mathcal{R}$ of Jacobson radical F-algebras which are not nil, satisfies the assumptions of Markov's theorem. Applying this theorem, one gets an example of a finitely generated F-algebra in which the nil and Jacobson radicals are distinct.

2.2. Köthe's nil ideal problem and some radicals of polynomial rings. As it was mentioned the Kostia's result described in the former subsection is related to

Köthe's nil ideal problem. Also some other Kostia's studies were related to this problem. In this section we present those concerning polynomial rings.

In [77] Krempa proved that Köthe's problem is equivalent to the problem whether polynomial rings in one indeterminate over nil rings are Jacobson radical. The Jacobson radical is contained in the Brown-McCoy radical. Thus attempting to get a positive solution of Köthe's problem it is reasonable to start with proving that the polynomial rings in one indeterminate over nil rings are Brown-McCoy radical. The problem whether it is true was open for several years. In [102] Puczyłowski and Smoktunowicz proved that it is indeed the case. This result can be viewed as a kind of an approximation of a positive solution of Köthe's problem. One can attempt to improve this approximation by replacing the Brown-McCoy radical with some other radicals laying between the Jacobson and Brown-McCoy radicals. One can also ask whether the same can be obtained for polynomial rings in sets of commuting or non-commuting indeterminates [99] or, more generally, for a description of the Brown-McCoy radical of polynomial rings in sets of commuting or non-commuting indeterminates. In [102] it was proved that the Brown-McCoy radical of the polynomial ring $R[x]$ in one indeterminate is equal $T[x]$, where $T = \cap\{I \lhd R \mid R/I$ is a prime ring with essential center$\}$. Applying this description one can easily show [100] that if A is a ring which is radical with respect to the smallest strong radical containing the nil radical, then $A[x]$ is Brown-McCoy radical.

In [31] Beidar, Fong and Puczyłowski proved that the approximation of Köthe's problem obtained in [102] can be extended to the Behrens radical. Recall that the Behrens radical is defined as the upper radical determined by the class of subdirectly irreducible rings whose hearts contain nonzero idempotents.

Theorem 2.2.1 ([31]). *Let A be a ring and n a positive integer. Suppose that for every natural number m and arbitrary left ideal L of the ring $M_m(A)$ of $m \times m$-matrices over A, the ring $L[x_1, \dots, x_n]$ in commuting indeterminates $x_1, \dots, x_n$ is Brown-McCoy radical. Then $A[x_1, \dots, x_n]$ is Behrens radical.*

Clearly this results extends to arbitrary set of commuting indeterminates.

If A is a nil ring, then for every m, $M_m(A)$ belongs to the smallest strong radical containing the nil radical. Hence Theorem 2.2.1 and the quoted above result from [100] imply that for every nil ring A, $A[x]$ is Behrens radical. Recently this result was sharply extended by Smoktunowicz [114]. She proved that if A is a nil ring, then primitive ideals of $A[x]$ are of the form $I[x]$.

In [31] an example of a Behrens radical ring A such that $M_n(A)$ for some n is not Behrens radical was constructed. This is rather surprising as matrix rings over

Brown-McCoy radical rings are Brown-McCoy radical and *the Behrens radical is the largest left (right) hereditary radical contained in the Brown-McCoy radical* [104].

The question *whether polynomial rings of several commuting or non-commuting indeterminates are over nil rings Brown-McCoy radical*, is still open. This problem was, in particular, studied by Ferrero and Wisbauer in [67], where it was proved that the questions for infinitely many commuting and non-commuting indeterminates are equivalent. In the last years Beidar was very much interested in this problem and related questions (in particular, *does there exist a prime ring with zero center whose central closure is a simple ring with unity element?*).

In [40] Beidar, Puczyłowski and Wiegandt studied some other approximations of Köthe's problem and related matters. The most interesting result related to Köthe's problem which was obtained there says that *no uniformly strongly prime ring contains nonzero left nil ideals*. Recall that a ring A is called *uniformly strongly prime* if it contains a finite set F such that for arbitrary $a, b \in A \setminus \{0\}$, $aFb \neq 0$.

As we have seen the problem of describing the Jacobson radical of polynomial rings in commuting indeterminates is difficult and strictly connected with Koethe's problem. It is much easier to describe the Jacobson radical of polynomial rings in non-commuting indeterminates. Namely [92] the Jacobson radical of the polynomial ring $A\{X\}$ in a set X of at least two non-commuting indeterminates is locally nilpotent and is equal $\mathcal{L}(A)\{X\}$, where $\mathcal{L}(A)$ is the locally nilpotent radical of A. Related are some recent results on the Jacobson radical of monomial algebras.

Recall that *monomial algebras* are defined as the factor algebras $F\{X\}/I$, where $F\{X\}$ is the polynomial algebra over a field F in a set X of non-commuting indeterminates and I is an ideal of $F\{X\}$ generated by monomials.

In [88, 89] Okninski asked: *is the Jacobson radical of monomial algebras locally nilpotent?* He proved that in the characteristic zero case the answer is positive if and only if the Jacobson radical of the monomial algebra $F\{X\}/I$ regarded as an algebra graded in the natural way by the free group generated by X is homogeneous. Homogeneity of the Jacobson radical of monomial algebras was proved in [72]. These answered the question for monomial algebras of characteristic zero. Next in [50] Belov and Gateva-Ivanova proved that the Jacobson radical of monomial algebras is nil. Finally in [29] Beidar and Fong answered Okninski's question in the positive.

Kostia dealt also with the structure of nil rings. In Beidar and Trokanová-Salavová [44] the structure of nil rings with descending chain condition on principal right ideals (briefly *MHR-rings*) was described as follows.

Theorem 2.2.2. *Let γ be the lower radical determined by the class of all zero-rings of prime order. For a ring A the following conditions are equivalent:*

  (i) *A is a nil MHR-ring,*

 (ii) *A is left T-nilpotent (that is, for any sequence of elements a_i of A, there exists an n such that $a_1 a_2 \ldots a_n = 0$) and $A \in \gamma$,*

(iii) *A is left T-nilpotent and every element of A has finite additive order.*

Theorem 2.2.2 and its consequences answer the problems raised by F. A. Szász [122].

2.3. On extensions of reduced rings and domains. Recall that a ring is called *reduced* if it contains no nonzero nilpotent elements.

Reduced rings play a very substantial role in ring theory. They were studied by many authors in quite different contexts. One of the main results in the area is due to Andrunakievich and Ryabukhin [5]. It states that *minimal prime ideals of reduced rings are completely prime*. As a consequence of this result one easily gets that *a ring is reduced if and only if it is a subdirect sum of domains*. This result was obtained independently by Bell [49] and later rediscovered and extended by many other authors.

It is well known that if $I \lhd \cdot A$, then I is a reduced ring (domain) if and only if so is A. The problem when I is a left (right) essential ideal of A is much more complicated. It appeared in [34] in studies of subhereditary radicals and in another context in [12]. Studies of the problem were continued in [32]. Below we present the main results on the subject obtained by Beidar, Ke and Puczyłowski in [34] and by Beidar, Fong and Puczyłowski in [32].

If I is a left (right) essential ideal of A, then we also say that A is a *left (right) essential extension of I.*

The fact that the classes of reduced rings and domains are not closed under left (right) essential extensions follows immediately from Theorem 1.5.8. Other examples to show that were constructed in [12, 103].

Theorem 2.3.1 ([34]).   (i) *If $\mathcal{K}$ is a hereditary class of reduced rings which is closed under left (right) essential extensions and subdirect products, then all rings in $\mathcal{K}$ satisfy a common polynomial identity.*

(ii) *The class $\mathcal{P}$ of all reduced rings which satisfy polynomial identity (not necessarily common) is closed under left (right) essential extensions.*

Theorem 2.3.2 ([34]). *The largest left hereditary class of reduced rings, which is closed under right essential extensions, coincides with the class $\mathcal{C}$ of all reduced rings A such that if B_1 and B_2 are left ideals of A and $B_1 \cap B_2 = 0$, then $B_1 B_2 = 0$.*

A domain belongs to the class $\mathcal{C}$ from Theorem 2.3.2 if and only if it is a left Ore domain. It is easy to see that each right essential extension of a domain is a prime ring and that prime rings are reduced if and only if they are domains. These and Theorem 2.3.2 give

Theorem 2.3.3 ([34]). *Every right essential extension of a left Ore domain is a domain.*

The class of all left essential extensions of domains will be denoted by $\mathcal{D}_e$ and the class of all left essential extensions of reduced rings will be denoted by $\mathcal{R}_e$.

The following theorem gives an analog of the quoted above result saying that reduced rings are subdirect sums of domains.

Theorem 2.3.4 ([32]). *Every ring in $\mathcal{R}_e$ is a subdirect sum of rings from $\mathcal{D}_e$.*

This theorem is a consequence of the following more general result.

Theorem 2.3.5 ([32]). *Suppose that a ring A is a left essential extension of a reduced ring R. Let $\mathrm{spec}(R)$ be the set of all minimal prime ideals of R and let*

$$\mathcal{P} = \{Q \mid Q \text{ is a prime ideal of } A \text{ and } Q \cap R \in \mathrm{spec}(R)\}.$$

Then

    (i) *For every $P \in \mathrm{spec}(R)$, $P = A^1 P = P A^1 \cap R$. In particular P is a left ideal of A;*

    (ii) *The map $\varphi : \mathcal{P} \to \mathrm{spec}(R)$, $Q \mapsto Q \cap R$, is a bijection;*

    (iii) *$\mathcal{P}$ is the set of all minimal prime ideals of A which do not contain R;*

    (iv) *$\bigcap\limits_{Q \in \mathcal{P}} Q = 0$.*

It is bit surprising that, as it was shown in [32], Theorem 2.3.4 cannot be inverted (clearly rings which are subdirect sums of domains are reduced).

2.4. Radicals induced by the total of rings. Following Kasch [73] the total $\mathrm{Tot}(A)$ of a ring A may be defined as the set

$$\mathrm{Tot}(A) = \{a \in A \mid aA \text{ does not contain nonzero idempotents}\}.$$

The total reminds to a radical, though the shortcoming is that $\mathrm{Tot}(A)$ need not be an ideal because it is not closed under addition. One can overcome this difficulty, and consider the upper radical of the class of rings with zero total.

Theorem 2.4.1 (Beidar and Wiegandt [48]). *The class*

$$\mathcal{M} = \{A \mid \mathrm{Tot}(A) = 0\}$$

is a weakly special class. The Kasch radical $\mathcal{K} = \mathcal{U}\mathcal{M}$ *is a supernilpotent radical which is left strong and left hereditary and hence is an N-radical.*

Also the *special radical* $\mathcal{K}_p$ may be defined as the upper radical $\mathcal{K}_p = \mathcal{U}\mathcal{M}_p$ where

$$\mathcal{M}_p = \{A \mid \mathrm{Tot}(A) = 0 \text{ and } A \text{ is a prime ring}\}.$$

Obviously $\mathcal{K} \subseteq \mathcal{K}_p$.

Theorem 2.4.2 ([48]). *The radical $\mathcal{K}_p$ is a special N-radical.*

Are $\mathcal{K}$ and $\mathcal{K}_p$ really different radicals?

Theorem 2.4.3 ([48]). *The Kasch radical $\mathcal{K}$ is a supernilpotent non-special radical, and so $\mathcal{K} \neq \mathcal{K}_p$.*

In the proof of Theorem 2.4.3 the construction of a ring $G \in \mathcal{K}_p \setminus \mathcal{K}$ is crucial. Kostia constructed such a ring G for which $\mathrm{Tot}(G) = 0$ but $\mathrm{Tot}(G/P) \neq 0$ for every prime ideal P of G. In the sequel this construction will be sketched.

Let U and V be the complete and discrete direct sum of countably infinite copies of the two element field, respectively, and put $B = U/V$. Let T denote the set of all ultrafilters of B,

$$T = \{\tau = B \setminus M \mid M \text{ is a maximal ideal in } B\}.$$

For each $e \in B$ we set

$$T(e) = \{\tau \in T \mid e \in \tau\}.$$

It is known that T is a compact Hausdorff space with a base of closed-open subsets

$$\{T(e) \mid e \in B\}.$$

Further, let

$$R = \mathbb{Q}\{x_\tau, y_\tau \mid \tau \in T\}$$

be the Weyl algebra in x_τ, y_τ ($\tau \in T$), that is the algebra over the rationals $\mathbb{Q}$ generated by x_τ, y_τ ($\tau \in T$), and subject to the relations of the commutators

$$[x_\tau, x_\sigma] = 0 = [y_\tau, y_\sigma], \qquad [x_\tau, y_\sigma] = \begin{cases} 1 & \text{if } \tau = \sigma \\ 0 & \text{otherwise} \end{cases}$$

for all $\tau, \sigma \in T$. Given a subset $S \subset T$, let R_S be the subalgebra

$$R_S = \mathbb{Q}\{x_\tau, y_\tau \mid \tau \in T \setminus S\}.$$

Both R and R_S are simple Ore domains, so they possess classical division rings of quotients Δ and Δ_S, respectively. Also

$$A_S = \Delta_S\{x_\tau, y_\tau \mid \tau \in S\}$$

is a simple Ore domain for every $S \subseteq T$.

Let us endow Δ with the discrete topology and let H be the $\mathbb{Q}$-algebra of all continuous functions $f : T \to \Delta$. Since T is compact, f continuous and Δ discrete, we have that $|f(T)| < \infty$. A finite subset $\{e_1, \ldots, e_n\}$ of B is said to be a partition of 1 if $e_1 + \cdots + e_n = 1$ and $e_i e_j = 0$ for all $i \neq j$. Since $|f(T)| < \infty$, there exists a partition of 1 such that $f|_{T(e_i)}$ is constant. A function f is said to be special, if $f(T(e_i)) \subseteq A_{T(e_i)}$ for all $i = 1, \ldots, n$.

Let G be the subset of H consisting of all special functions. The subset G is then a subalgebra of H and G is a ring with the desired properties.

The Kasch radical $\mathcal{K}$ contains the Behrens radical $\mathcal{B}$, and $\mathcal{K}_p$ is contained in the upper radical of all von Neumann regular rings. $\mathcal{K}$, as well as $\mathcal{K}_p$, is not comparable with the Brown-McCoy radical.

Semiprime rings with bounded index of nilpotent elements were investigated by Beidar and Mikhalev [38]. Beidar [24] continued these investigations, and in particular, for prime rings he got

Theorem 2.4.4 ([24]). *The following are equivalent:*

   (i) *A is a prime ring with bounded index n of nilpotent elements and* $\mathrm{Tot}(A) = 0$;

   (ii) *A is isomorphic to the matrix ring $M_n(D)$ over a division ring D.*

In [24] he proved that for a ring with zero total and bounded indices of nilpotent elements, the maximal right ring of quotients equals the maximal left ring of quotients, and is isomorphic to a finite direct sum of matrix rings over strongly regular selfinjective rings.

3. Rings with involution

3.1. Involution rings with dcc on ∗-biideals. A ring A is said to be an *involution ring*, if there is defined a unary operation $* : a \longmapsto a^*$, called *involution*, subject to the familiar identities

$$(a + b)^* = a^* + b^*, \qquad (ab)^* = b^* a^*, \qquad a^{**} = a$$

for all $a, b \in A$. Rings with involution have been studied mostly as rings with an additional operation (cf. for instance [37]). Far less attention has been payed to the consistent study of involution rings as objects of the category of involution rings with mapping preserving also involution. In this category life is more diffi-cult. Proving structure theorems for rings, the notion of one-sided ideals (a link to module theory) plays a decisive role. A one-sided ideal closed under involution is a two-sided ideal, so in the category of involution rings one-sided ideals cannot be used. An appropriate notion seems to be that of ∗-biideals. A subring B of an involution ring A is a *∗-biideal*, if $BAB \subseteq B$ and $B^* = B$.

Loi [83] proved that a semiprime involution ring A satisfies *dcc* (descending chain condition) on principal ∗-biideals if and only if A, as a ring satisfies dcc on principal right ideals. Loi and Wiegandt [84] showed that for such an involution ring A the Baer and Jacobson radicals coincide and they are contained in the maxi-mal torsion ideal, the *torsion radical*, of A; moreover, every involution ring A with dcc on principal ∗-biideals splits into a direct sum of its torsion radical $\tau(A)$ and of a uniquely determined torsionfree ideal.

Dropping the attribute "principal", stronger results can be proved.

Theorem 3.1.1 (Beidar and Wiegandt [47]). *If an involution ring A has dcc on ∗-biideals, then its Jacobson radical $\mathcal{J}(A)$ satisfies dcc on additive subgroups, and hence $\mathcal{J}(A)$ is nilpotent.*

Notice that dcc on ∗-biideals is a much stronger condition that dcc on one-sided ideals: the Jacobson radical of an artinian ring need not be artinian.

Corollary 3.1.2 ([47]). *An involution ring A satisfies dcc on ∗-biideals if and only if A, as a ring, is artinian with artinian Jacobson radical.*

Thus the structure of involution rings with dcc on ∗-biideals is fully described (cf. Kertész and Widiger [74]).

In the proof of Theorem 3.1.1—among others—a lemma was used: *if the Baer radical $\beta(A)$ of an involution ring A is infinite, then there exists an infinite ∗-ideal I of A such that $I^3 = 0$.* This lemma plays a crucial role in the topologization of commutative rings without involution (cf. Arnautov [13]). The proof given in [47] works for non-commutative, associative rings without involution, and so Beidar and Mikhalev [39] proved that *on every associative ring with infinite Baer radical there exists a nondiscrete topology.*

3.2. Involution rings with acc on ∗-biideals. The study of involution rings satisfying *acc* (ascending chain condition) on ∗-biideals requires even more involved techniques and gives surprising results.

Theorem 3.2.1 (Beidar and Wiegandt [47]). *If an involution ring A has acc on ∗-biideals, then the additive group of its Baer radical is finitely generated.*

The famous Hilbert Basis Theorem states that a ring A with unity element is noetherian if and only if the polynomial ring $A[x]$ is noetherian. Given an involution ring A, the involution on A extends to an involution on $A[x]$ by defining $x^* = x$ or $x^* = -x$. For involution rings the following counterpart of the Hilbert Basis Theorem can be proved.

Theorem 3.2.2 ([47]). *For an involution ring A the following conditions are equivalent:*

(i) *the polynomial ring $A[x]$ has acc on ∗-biideals,*

(ii) *the ring A is semiprime, finite and hence a finite direct sum of matrix rings over finite fields.*

Theorem 3.2.2 was proved in [47] for nonassociative rings, with the necessary modification of the definition of ∗-biideals and with an extra condition imposed on A in terms of fields of rational functions over prime fields.

The results of [47] were announced in [46].

3.3. Involution rings and primitivity. The involutive version of primitivity was introduced by Rowen [105] and he noted: "There does not seem to be a good ∗-analogue for the density theorem in general, although there is an excellent version for ∗-primitive rings having minimal left ideals." Nevertheless, ∗-primitive involution rings, as seen from the subsequent results, are worth for further scrutiny.

Given an involution ring R and an R-module, M, we say that M is *∗-faithful* if

$$\{r \in R \mid rM = 0 = r^*M\}.$$

The involution ring R is said to be *∗-primitive* if there is a simple ∗-faithful R-module.

On the direct sum D of two anti-isomorphic rings S and $T = S^{\mathrm{op}}$ (without involution), an involution $*$ may be defined, called the *exchange involution*, as follows:

$$(s, t)^* = (t, s) \quad \text{for all } s \in S \text{ and } t \in T.$$

Let an involution ring R be a subdirect sum of S and T, and $\psi : S \to T$ the anti-isomorphism between S and T. We say that the involution $*$ on R is of *exchange type* if

$$(s, t)^* = (\psi^{-1}(t), \psi(s))$$

for all $(s, t) \in S \underset{\text{subd.}}{\boxplus} T$.

The next result is the involutive version of the density theorem for rings.

Theorem 3.3.1 ([36]). *An involution ring R is $*$-primitive if and only if either*

(I) *R is left primitive (and then also right primitive), or*

(II) *R is a subdirect sum of two anti-isomorphic (left, resp. right) primitive rings with exchange type involution; or equivalently, R is an essential extension of the direct sum of two anti-isomorphic (left resp. right) primitive rings endowed with the exchange involution.*

One may have the feeling that in case (I) of Theorem 3.3.1 the left and right primitive ring R has a row and column finite matrix representation. This is not always the case, as shown by Kostia (January 26, 2004).

Theorem 3.3.2 ([36]). *Let V be a separable Hilbert space over the complex number field C with inner product $\langle \, , \, \rangle$ and let H be the C^*-algebra of bounded linear operators on V with involution $*$ such that $\langle Ax, y \rangle = \langle x, A^*x \rangle$ for all $x, y \in V$ and $A \in H$. Then V is a simple faithful left H-module with $\mathrm{End}(_H V) = C$ and for any (algebraic) basis $E = \{e_i \mid i \in I\}$ of V over C there exists an operator $A \in H$ such that the matrix of A relative to the basis E is not row finite.*

Since one-sided ideals make no sense in the category of involution rings, it is desirable to involve $*$-biideals into the description of $*$-primitive rings.

Theorem 3.3.3 ([36]). *An involution ring R is $*$-primitive if and only if $R^2 \neq 0$ and R contains a maximal $*$-biideal which does not contain nonzero $*$-ideals of R.*

The alien notion "left ideal" can be expelled from the description of $*$-primitive and even of $*$-prime involution rings. An involution ring R is said to be $*$-*prime* if for any two $*$-ideals K and L of R, from $KL = 0$ it follows $K = 0$ or $L = 0$.

Theorem 3.3.4 ([36]). *A ∗-prime involution ring has a minimal left ideal if and only if it has a minimal ∗-biideal. Every ∗-prime involution ring with a minimal ∗-biideal is ∗-primitive.*

4. Nonassociative rings

4.1. Sufficient conditions for a well-behaved radical theory.

Moving from the variety of associative rings towards the variety of nonassociative rings, the radical theory degenerates. For instance, for associative rings every semisimple class is hereditary, but for nonassociative ring the semisimple class $\mathcal{S}\gamma$ of a radical class γ is hereditary if and only if γ is an A-radical (that is, the radicality depends only on the additive group: $A \in \gamma$ and $A^+ \cong B^+ \Rightarrow B \in \gamma$). Successive attempts were made to formulate conditions on *universal classes* (that is, classes of rings being closed under taking homomorphic images and ideals) of nonassociative rings to ensure affirmative answer for three major issues of general radical theory (see Krempa and Terlikowska [78], Puczyłowski [94], [98], Terlikowska-Osłowska [123], [124] and Veldsman [127], [128]):

i) the validity of the ADS-*Theorem*: for every radical γ, $I \triangleleft A$ implies $\gamma(I) \triangleleft A$ for all rings A and $I \triangleleft A$;

ii) The validity of *Sands' Theorem*: σ is a semisimple class if and only if σ is *regular*, σ has the *coinductive property* (if $I_1 \supseteq \cdots \supseteq I_\lambda \supseteq \ldots$ is a descending chain of ideals of a ring A such that $A/I_\lambda \in \sigma$ for all λ, then also $A/\bigcap_\lambda I_\lambda \in \sigma$), and σ is closed under extensions ($I \triangleleft A$, $I \in \sigma$ and $A/I \in \sigma$ imply $A \in \sigma$);

iii) the termination of the lower radical construction at the first limit ordinal (see Section 1.1).

Beidar [21] accomplished this project in a different way; he required purely nonassociative ring-theoretical conditions and obtained the same results.

For a subset B of a ring A the ideal of A generated by B will be denoted by $\langle B \rangle_A$. For any ring A we define

$$A^{(1)} = A \quad \text{and} \quad A^{(n+1)} = A^{(n)} \cdot A^{(n)}$$

for $n = 1, 2, \ldots$, that is $A^{(n+1)}$ is the set of all finite sums $\sum a_i b_i$ with $a_i, b_i \in A^{(n)}$.

For answering the three major issues of radical theory, Beidar considered the following conditions imposed on a universal class $\mathbb{A}$ of nonassociative rings:

(B1) to every $K \lhd I \lhd A \in \mathbb{A}$ and every element $a \in A$ there exists an integer $m > 0$ such that $aK^{(m)} + K^{(m)}a \subseteq K$,

(B2) to every $K \lhd I \lhd A \in \mathbb{A}$ and every element $a \in A$ there exists an integer $n > 0$ such that $(\langle aK + Ka + K \rangle_I)^{(n)} \subseteq K$,

(B3) to every $K \lhd I \lhd A \in \mathbb{A}$ there exists an integer $t > 0$ such that $\langle K^{(t)} \rangle_I \subseteq K^2$.

Theorem 4.1.1 (Nikitin [87] and Beidar [21]). *If the universal class $\mathbb{A}$ satisfies conditions* (B1) *and* (B2), *then the ADS-Theorem is valid.*

Theorem 4.1.2 ([21]). *If the universal class $\mathbb{A}$ satisfies conditions* (B1), (B2) *and* (B3), *then Sands' Theorem is valid.*

Theorem 4.1.3 ([21]). *Let $\mathbb{A}$ be a universal class satisfying conditions* (B1) *and* (B2). *If in* (B2) *the integer n depends only on the ring A, then the lower radical construction terminates at most at the first limit ordinal.*

Examples for universal classes are the varieties of alternative rings, of Jordan R-algebras with $1/2 \in R$ and Andrunakievich s-varieties. This latter notion was introduced by Anderson and Gardner [2] who gave several examples for that notion (for instance, 4-permutable rings and autodistributive rings).

In [33] Beidar, Glavatskiǐ and Mikhalev proved analogous results to Theorems 4.1.2 and 4.1.3 for nonassociative topological algebras under conditions (B1), (B2), and (B3) demanding in (B3) that the topological closure of $\langle K^{(t)} \rangle_I$ be contained in the topological closure of K^2.

4.2. The splitting of the torsion radical. In the study of algebraic structures the main goal is to prove structure theorems. Fairly good structure theorems are the decompositions into a direct sum. Direct decompositions into two components of diverse properties can be considered as a first but basic step in implementing this project. In this section we shall deal with the problem when the torsion radical of an alternative ring is a direct summand.

In every ring A the set of all additively torsion elements forms an ideal $\tau(A)$. The assignment $A \to \tau(A)$ is a radical assignment, and $\tau(A)$ is called the *torsion radical* of A. Notice that the torsion radical τ is an A-*radical* (cf. Sections 3.1 and 4.1). F. Szász [121] proved the splitting of the torsion radical in artinian associative rings, a result which has become folklore: *every artinian associative ring A decomposes into the direct sum $A = \tau(A) \oplus F$ of its torsion radical $\tau(A)$ and of a uniquely determined torsionfree ideal F.* Let us observe that, although in direct decomposition theorems the components are determined usually only up to isomorphisms, here both direct summands are uniquely determined ideals.

Szász' Theorem was extended to associative rings with dcc on principal right ideals by Christine Ayoub [16] and Dinh Van Huynh [55] simultaneously by different methods.

A ring A is said to be *alternative* if A satisfies

$$a^2b = a(ab) \quad \text{and} \quad (ab)b = ab^2$$

for all $a, b \in A$. Alternative rings are obviously nonassociative generalizations of associative rings.

Widiger [131] proved Szász' Theorem for artinian alternative rings. It was conjectured that Szász' Theorem must be true also for alternative rings with dcc on principal right ideals. Analyzing the previous proofs and developing new methods Beidar and Wiegandt [45] proved

Theorem 4.2.1. *Every alternative ring A with dcc on principal right ideals is the direct sum of its torsion radical $\tau(A)$ and of a uniquely determined torsionfree ideal F.*

The proof consists of two parts. Following the ideas of Ayoub [16], it was proved that for any nonassociative ring A with dcc on principal right ideals the following conditions are equivalent:

  (i) the torsion radical $\tau(A)$ is a direct summand,
  (ii) D^2 is torsionfree where D denotes the maximal divisible ideal of A,
  (iii) if a subdirectly irreducible factor ring B of A has a torsion heart then B is a torsion ring.

Ayoub [16] proved the validity of (ii) in the associative case. Dealing with alternative rings with dcc on principal right ideals, in [45] the validity of (iii) was proved.

As a by-product, a splitting theorem for Jordan rings was obtained in [45]. A ring is a *Jordan ring*, if it is commutative and satisfies the identity $(a^2b)a = a^2(ba)$. Suppose that A is a Jordan ring such that its additive group A^+ is the direct sum of a reduced torsion subgroup B^+ and its maximal divisible subgroup D^+. If for any finitely many elements $x_1, \ldots, x_n \in A/\tau(A)$ there exists an element $e \in A/\tau(A)$ such that $x_ie = x_i = x_ie^2, i = 1, \ldots, n$, then D^2 is torsionfree. Hence an application of the equivalence of (i) and (ii) gives

Theorem 4.2.2 ([45]). *If a Jordan ring satisfies the above requirements, then its torsion radical splits off.*

The splitting of the torsion radical in abstract affine near-rings with dcc on principal right ideals was proved by Birkenmeier and Wiegandt [54].

References

[1] Amitsur, S. A. *Algebras over infinite fields,* Canad. J. Math. **8** (1956), 355–361.

[2] Anderson T.; Gardner, B. J. *Semisimple classes in a variety satisfying an Andrunakievich Lemma,* Bull. Austral. Math. Soc. **18** (1978), 187–200.

[3] Andrunakievich, V. A. *Radicals of associative rings I,* (Russian), Mat. Sb. **44** (1958), 179–212; English trasnl.: Amer. Math. Soc. Transl. **52** (1966), 95-128.

[4] Andrunakievich, V. A. *Radicals of associative rings II,* (Russian), Mat. Sb. **55** (1961), 329–346; English transl.: *Amer. Math. Soc. Transl.* (2) **52** (1966), 129–150.

[5] Andrunakievich, V. A.; Ryabukhin, Yu. M. *Rings without nilpotent elements and completely prime ideals,* (Russian) Dokl. Akad. Nauk SSSR **180** (1968), 9–11.

[6] Andrunakievich, V. A.; Ryabukhin, Yu. M. Radicals of Algebras and Structure Theory, Nauka, Moscow, 1979 (in Russian).

[7] Andruszkiewicz, R. R. *On accessible subrings of associative rings,* Proc. Edinburgh Math. Soc. **35** (1992), 101–107.

[8] Andruszkiewicz, R. R. *Essential cover and essential closure,* Serdica Math. J. **30** (2004), 505–512.

[9] Andruszkiewicz, R. R.; Puczyłowski, E. R. *Kurosh's chains of associative rings,* Glasgow Math. J. **32** (1990), 67–69.

[10] Andruszkiewicz, R. R.; Puczyłowski, E. R. *Accessible subrings and Kurosh's chains of associative rings,* Algebra Colloq. **4** (1997), 79–88.

[11] Armendariz, E. P.; Leavitt, W. G. *The heredity property in the lower radical construction,* Canad. J. Math. **20** (1968), 474–476.

[12] Armendariz, E. P.; Birkenmeier, G. F.; Park, J. K. *Rings Containing Ideals with Bounded Index.* Comm. Algebra **30** (2002), 787–801.

[13] Arnautov, V. I. *Nondiscrete topologization of infinite commutative rings,* (Russian), Mat. Issled. Kishinev **5** (1970), 3–15.

[14] Arnautov, V. I.; Beidar, K. I.; Glavatskiĭ, S. T.; Mikhalev, A. V. *The intersection property in the theory of radicals of topological algebras,* (Russian), Trudy Sem. Petrovsk. **15** (1991), 178–188.

[15] Arnautov, V. I.; Beidar, K. I.; Glavatskiĭ, S. T.; Mikhalev, A. V., *The intersection property in the theory of radicals of topological algebras,* Contemporary Math. **131** (1992), 205–225.

[16] Ayoub, Chr. *Conditions for a ring to be fissile,* Acta Math. Acad. Sci. Hungar. **30** (1977), 233–237.

[17] Beidar, K. I. *On radicals of finitely generated algebras,* (Russian), Russian Math. Surveys **36** (1981), 203–204.

[18] Beidar, K. I. *A chain of Kurosh may have an arbitrary finite length,* Czechosl. Math. J. **32** (1982), 418–422.

[19] Beidar, K. I. *Examples of rings and radicals,* in "Radical Theory", Proceeding of the Conference at Eger, 1982, pp. 19–46, Colloq. Math. Soc. J. Bolyai, Vol. 38. North-Holland, Amsterdam, 1985.

[20] Beidar, K. I. *Atoms in the "lattice" of radicals,* (Russian), Mat. Issled. Kishinev **85** (1985), 21–31.

[21] Beidar, K. I. *Semisimple classes and the lower radical,* (Russian), Mat. Issled. Kishinev **105** (1988), 13–29.

[22] Beidar, K. I. *The intersection property for radicals,* (Russian), Usp. Mat. Nauk **44** (1989), 1 (265), 187–188.

[23] Beidar, K. I. *On questions of B. J. Gardner and A. D. Sands,* J. Austral. Math. Soc. **56** (1994), 314–319.

[24] Beidar, K. I. *On rings with zero total,* Beiträge Alg. und Geom. **38** (1997), 233–239.

[25] Beidar, K. I. *On principally hereditary radicals,* Comm. Algebra **26** (1998), 3899–3912.

[26] Beidar, K. I. *On essential extensions, maximal extensions and iterated maximal essential extensions in radical theory,* in "Radical Theory", Proceedings of the Conference at Szekszárd, 1991, pp. 17–26, Colloq. Math. Soc. J. Bolyai, Vol. 61. North-Holland, Amsterdam, 1993.

[27] Beidar, K. I.; Fong, Y.; Ke, W.-F.; Shum, K. P. *On radicals with semisimple essential covers.* Preprint, 1995.

[28] Beidar, K. I.; Fong, Y.; Wang, C. S. *On the lattice of strong radicals,* J. Algebra **180** (1996), 334–340.

[29] Beidar, K. I.; Fong, Y. *On radicals of monomial algebras,* Comm. Algebra **26** (1998), 3913–3919.

[30] Beidar, K. I.; Fong, Y.; Ke, W.-F. *On complemented radicals,* J. Algebra **201** (1998), 328–356.

[31] Beidar, K. I.; Fong, Y.; Puczyłowski, E. R. *Polynomial rings over nil rings cannot be homomorphically mapped onto rings with nonzero idempotents,* J. Algebra **238** (2001), 389–399.

[32] Beidar, K. I.; Fong, Y.; Puczyłowski, E. R. *On essential extensions of reduced rings and domains,* Arch. Math. **83** (2004), 344–352.

[33] Beidar, K. I.; Glavatskiǐ, S. T.; Mikhalev, A. V. *Semisimple classes and lower radicals of topological non-associative algebras,* (Russian), Trudy Sem. Petrovsk. **14** (1989), 250–261; English transl.: J. Soviet Math., **51** (1990), 2487–2496.

[34] Beidar, K. I.; Ke, W.-F.; Puczyłowski, E. R. *On subhereditary radicals and reduced rings,* Proc. Royal Soc. Edinburgh Sec. A **132** (2002), 255–266.

[35] Beidar, K. I.; Ke, W.-F.; Puczyłowski, E. R. *On matrix rings and subhereditary radicals,* Comm. Algebra **32** (2004), 2827–2839.

[36] Beidar, K. I.; Márki, L.; Mlitz, R.; Wiegandt, R. *Primitive involution rings,* Acta Math. Hungar. **109** (2005), 357–368.

[37] Beidar, K. I.; Martindale, W. S., III; Mikhalev, A. V. Rings with generalized identities, Marcel Dekker, New York, 1996.

[38] Beidar, K. I.; Mikhalev, A. V. *Semiprime rings with bounded index of nilpotent elements,* (Russian), Trudy Sem. Petrovsk. **13** (1988), 237–249; English transl.: J. Soviet Math. **50**(1) (1990), 1518–1526.

[39] Beidar, K. I.; Mikhalev, A. V. *On topologization of rings with infinite lower Baer radical,* Contemporary Mathematics **184** (1995), 43–47.

[40] Beidar, K. I.; Puczyłowski, E. R.; Wiegandt, R. *Radicals and polynomial rings,* J. Austral. Math. Soc. **72** (2000), 23–31.

[41] Beidar, K. I.; Salavová, K. *Some examples of supernilpotent nonspecial radicals,* Acta Math. Hungar. **40** (1982), 109–112.

[42] Beidar, K. I.; Salavová, K. *The lattice of N-radicals, left strong radicals, and left hereditary radicals,* (Russian), *Acta Math. Hungar.* **42** (1983), 81–95.

[43] Beidar, K. I.; Salavová, K. *Additive radicals,* Czechosl. Math. J. **39** (1989), 659–673.

[44] Beidar, K. I.; Trokanová-Salavová, K. *On nil rings satisfying minimum condition on principal right ideals,* Acta Math. Hungar. **55** (1990), 197–200.

[45] Beidar, K. I.; Wiegandt, R. *Splitting theorems for nonassociative rings,* Publ. Math. Debrecen **38** (1991), 121–143.

[46] Beidar, K. I.; Wiegandt, R. *Rings with involution and chain conditions on biideals,* (Russian), *Usp. Mat. Nauk* **485** (293) (1993), 159–160.

[47] Beidar, K. I.; Wiegandt, R. *Rings with involution and chain conditions,* J. Pure & Appl. Algebra **87** (1993), 205–220.

[48] Beidar, K. I.; Wiegandt, R. *Radicals induced by the total of rings,* Beiträge Alg. und Geom. **38** (1997), 149–159.

[49] Bell, H. E. *Duo rings, some applications to commutativity theorems,* Canad. Math. Bull. **11** (1968), 375–380.

[50] Belov, A.; Gateva-Ivanova, T. *Radicals of monomial algebras,* in Proceedings of First International Tainan-Moscow Algebra Workshop, (Tainan, 1994), 159–169, de Gruyter, Berlin-New York, 1996.

[51] Birkenmeier, G. F. *Rings which are essentially supernilpotent,* Comm. Algebra **22** (1994), 1063–1082.

[52] Birkenmeier, G. F. *Radicals whose essential covers are semisimple classes,* Comm. Algebra 22 (1994), 6239–6258.

[53] Birkenmeier, G. F.; Wiegandt, R. *Essential covers and complements of radicals,* Bull. Austral Math. Soc. **53** (1996), 261–266.

[54] Birkenmeier, G. F.; Wiegandt, R. *Supplementing radicals and decompositions of near-rings,* Acta Math. Hungar. **94** (2002), 269–280.

[55] Van Huynh, D. *Die Spaltbarkeit von MHR-Ringen,* Bull. Acad. Polon. Sci. **25** (1977), 939–941.

[56] Divinsky, N.; Krempa, J.; Sulinski, A. *Strong radical properties of associative and alternative rings,* J. Algebra **17** (1971), 369–388.

[57] Dubrovin, N. I. *Chain domains,* (Russian), Vestnik Moskov. Univ. Ser. Mat. Meh. **2** (1980), 51–54.

[58] France-Jackson, H. *On prime essential rings,* Bull. Austral. Math. Soc. **47** (1993), 287–290.

[59] Gardner, B. J. *A note on radicals and polynomial rings,* Math. Scand. **31** (1972), 83–88.

[60] Gardner, B. J. *Absolute retracts, relatively injective simple objects and Brown-McCoy radicals,* Contemp. Math. **9** (1982), 227–283.

[61] Gardner, B. J. *Simple rings whose lower radicals are atoms,* Acta Math. Hungar. **43** (1984), 131–135.

[62] Gardner, B. J. *Problem,* in "Rings, modules and radicals", Proc. Conf. Hobart, 1987, Pitman Res. Notes in Math. 204, Longman Sci. & Tech., 1989, p. 191.

[63] Gardner, B. J.; Stewart, P. N. *Prime essential rings,* Proc. Edinburgh Math. Soc. **34** (1991), 241–250.

[64] Gardner, B. J.; Stewart, P. N. *On semisimple radical classes,* Bull. Austral. Math. Soc. **13** (1975), 349–353.

[65] Gardner, B. J.; Wiegandt, R. Radical theory of rings, Marcel Dekker, New York, 2004.

[66] Guo, J.-Y. *On the termination of the construction of the lower radical class for a class of associative rings,* (Chinese), Chinese Ann. Math. Ser. A **8** (1987), 433–444.

[67] Ferrero, M.; Wisbauer, R. *Unitary strongly prime rings and related radicals,* J. Pure Appl. Algebra **181** (2003), 209–226.

[68] Heinicke, A. G. *A note on lower radical constructions for associative rings,* Canad. Math. Bull. **11** (1968), 23–30.

[69] Hendelman, D.; Lawrence, J. *Strongly prime rings,* Trans. Amer. Math. Soc. **211** (1975), 209–223.

[70] Hoffman, A. E.; Leavitt, W. G. *A note on the termination of the lower radical construction,* J. London Math. Soc. **43** (1968), 617–618.

[71] Jaegermann, M.; Sands, A. D. *On normal radicals, N-radicals, and A-radicals,* J. Algebra **50** (1978), 337–349.

[72] Jespers, E.; Puczyłowski, E. R. *The Jacobson and Brown-McCoy radicals of rings graded by free groups,* Comm. Algebra **19** (1991), 551–558.

[73] Kasch, F. Partiell invertierbare Homomorphismen und das Total, Algebra Berichte 60, Reinhard Fisher, München, 1988.

[74] Kertész, A.; Widiger, Λ. *Artinsche Ringe mit artinschem Radikal,* J. reine angew. Math. **242** (1970), 8–15.

[75] Korolczuk, H. *Lattices of radicals of associative rings,* (Polish), unpublished.

[76] Köthe, G. *Die Struktur der Ringe, deren Restklassenring nach dem Radikal vollständig reduzibel ist,* Math. Zeit. **32** (1930), 161–186.

[77] Krempa, J. *Logical connections among some open problems in non-commutative rings,* Fund. Math. **76** (1972), 121–130.

[78] Krempa, J.; Terlikowska, B. *Theory of radicals in self-dual categories,* Bull. Acad. Polon. Sci. **22** (1974), 367–373.

[79] Leavitt, W. G. *A minimally embeddable ring,* Period. Math. Hungar. **12** (1981), 129–140.

[80] Leavitt, W. G.; Watters, J. F. *Special closure, M-radicals and relative complements,* Acta Math. Acad. Hungar. **28** (1976), 55–67.

[81] van Leeuwen, L. C. A.; Jenkins, T. L. *A supernilpotent non-special radical class,* Bull. Austral. Math. Soc. **9** (1973), 343–348.

[82] Loi, N. V. *Essentially closed radical classes,* J. Austral. Math. Soc., Ser. A **35** (1983), 132–142.

[83] Loi, N. V. *On the structure of semiprime involution rings,* in "General Alg.", Proc. Krems Conf. 1988, North-Holland, 1990, 155–161.

[84] Loi, N. V.; Wiegandt, R. *On involution rings with minimum condition,* Ring Theory, Israel Math. Conf. Proc. **1** (1989), 203–214.

[85] L'vov, I. A.; Sidorov, A. V. *On the stabilization of Kurosh chains,* (Russian), Mat. Zametki **36** (1984), 815–821.

[86] Markov, V. T. *Some examples of finitely generated algebras,* (Russian), Russian Math. Surveys **36** (1981), 185–186.

[87] Nikitin, A. A. *On heredity of radicals in Jordan rings,* (Russian), Alg. i Logika **17** (1978), 303–315.

[88] Okninski, J. *On monomial algebras,* Arch. Math. **50** (1988), 417–423.

[89] Okninski, J. Semigroup Algebras, Marcel Dekker, New York, 1991.

[90] Ortiz, A. H. *A construction in general radical theory,* Canad. J. Math. **22** (1970), 1097–1100.

[91] Osłowski, B.; Puczyłowski, E. R. *On strong radicals of alternative algebras,* Bull. Acad. Polon. Sci. **25** (1977), 845–850.

[92] Puczyłowski, E. R. *Radicals of polynomial rings, power series rings and tensor products,* Comm. Algebra **8** (1980), 1699–1709.

[93] Puczyłowski, E. R. *Hereditariness of strong and stable radicals,* Glasgow Math. J. **23** (1982), 85–90.

[94] Puczyłowski, E. R. *On semisimple classes of associative and alternative rings,* Proc. Edinburgh Math. Soc. **27** (1984), 1–5.

[95] Puczyłowski, E. R. *On Sands' questions concerning strong and hereditary radicals,* Glasgow Math. J. **28** (1986), 1–3.

[96] Puczyłowski, E. R. *On questions concerning strong radicals of associative rings,* Quaestiones Math. **10** (1987), 321–338.

[97] Puczyłowski, E. R. *On essential extensions of rings,* Bull. Austral. Math. Soc. **35** (1987), 379–386.

[98] Puczyłowski, E. R. *On general theory of radicals,* Alg. Universalis **30** (1993), 53–60.

[99] Puczyłowski, E. R. *Some questions concerning radicals of associative rings,* in "Radical Theory", Proceedings of the Conference at Szekszard, 1991, pp. 209–227, Colloq. Math. Soc. J. Bolyai, Vol. 61, Noth-Holland, Amsterdam, 1993.

[100] Puczyłowski, E. R. *Some results and questions on nil rings,* Mat. Contemp. **16** (1999), 265–280.

[101] Puczyłowski, E. R.; Roszkowska, E. *On atoms and coatoms of lattices of associative rings,* Comm. Algebra **20** (1992), 955–977.

[102] Puczyłowski, E. R.; Smoktunowicz, A. *On maximal ideals and the Brown-McCoy radical of polynomial rings,* Comm. Algebra **26** (1998), 2473–2482.

[103] Puczyłowski, E. R.; Zand, H. *Subhereditary radicals of associative rings,* Algebra Colloq. **6** (1999), 215–223.

[104] Puczyłowski, E. R.; Zand, H. *The Brown-McCoy radical and one-sided ideals,* Quaestiones Math. **19** (1996), 47–58.

[105] Rowen, L. H. Ring theory I, Academic Press, Boston, 1988.

[106] Rjabuhin, Ju. M. *Overnilpotent and special radicals.* (Russian) 1965 Studies in Algebra and Math. Anal. (Russian) pp. 65–72

[107] Ryabukhin, Yu. M. *Non-comparable nil radicals and non-special supernilpotent radicals.* (Russian) Alg. i Logika **14** (1975), 86–99.

[108] Sands, A. D. *Radicals and Morita contexts,* J. Algebra **24** (1973), 335–345.

[109] Sands, A. D. *On normal radicals,* J. London Math. Soc. **11** (1975), 361–365.

[110] Sands, A. D. *Strong upper radicals,* Quart. J. Math. Oxford **27** (1976), 21–24.

[111] Sands, A. D. *On relations among radical properties,* Glasgow Math. J. **18** (1977), 17–23.

[112] Sands, A. D. *Radical properties and one-sided ideals,* in "Contribution to General Algebra 4", Proceedings of the Conference at Krems, 1985, pp. 153–171, Verlag Holder-Pichler-Tempsky, Vienna, Verlag B. G. Teubner, Stuttgart, 1987.

[113] Sands, A. D. *Radicals and one-sided ideals,* Proc. Royal Soc. Edinburgh **103A** (1986), 241–251.

[114] Smoktunowicz, A. *On primitive ideals in polynomial rings over nil rings,* Algebr. Represent. Theory **8** (2005), 69–73.

[115] Sidorov, A. V. *On the stabilization of Kurosh chains in the class of semigroups with zero* (Russian), Sibir. Mat. Zh. **29** (1988), 131–136.

[116] Snider, R. L. *Complemented hereditary radicals,* Bull. Austral. Math. Soc. **4** (1971), 307–320.

[117] Snider, R. L. *Lattices of radicals,* Pacific J. Math. **40** (1972), 207–220.

[118] Stewart, P. N. *Semi-simple radical classes,* Pacific J. Math. **32** (1970), 249–254.

[119] Stewart, P. N. *On the lower radical construction,* Acta Math. Acad. Sci. Hungar. **25** (1974), 31–32.

[120] Suliński, A.; Anderson, T.; Divinsky, N. *Lower radical properties for associative and alternative rings,* J. London Math. Soc. **41** (1966), 417–424.

[121] Szász, F. *Über artinsche Ringe,* Bull. Acad. Polon. Sci. **11** (1963), 351–354.

[122] Szász, F. A. *Radikale der Ringe,* Deutscher Verlag d. Wiss., Berlin, 1975; English edition: Radicals of Rings, John Wiley & Sons, Chichester, 1981.

[123] Terlikowska-Osłowska, B. *Category with a self-dual set of axioms,* Bull. Acad. Polon. Sci. **25** (1977), 1207–1214.

[124] Terlikowska-Osłowska, B. *Radical and semisimple classes of objects in categories with self-dual set of axioms,* Bull. Acad. Polon. Sci. **26** (1978), 7–13.

[125] Tumurbat, S. *A note on normal radicals and principally left and right strong radicals,* Southeast Asian Bull. Math. **28** (2004), 363–368.

[126] Tumurbat, S.; Wiegandt, R. *On subhereditary radicals and Brown–McCoy semisimple rings,* Bul. Acad. Ştiinţe Rep. Moldova, Matematica **3**(34) (2000), 11–20.

[127] Veldsman, S. *Sufficient condition for a well-behaved Kurosh-Amitsur radical theory,* Proc. Edinburgh Math. Soc. **32** (1989), 377–394.

[128] Veldsman, S. *Sufficient condition for a well-behaved Kurosh-Amitsur radical theory II,* in "Rings, Modules and Radicals, Proc. Conf. Hobart 1987", Pitman Res. Notes in Math. 204, Longman Sci. & Tech., 1989, 142–152.

[129] Watters, J. F. *Lower radicals in associative rings,* Canad. J. Math. **21** (1969), 466–476.

[130] Watters, J. F. *On the lower radical construction for algebras,* Preprint, 1985.

[131] Widiger, A. *Decomposition of artinian alternative rings,* Beiträge Alg. und Geom., Halle DDR **12** (1982), 57–71.

[132] Wu, T.-S. *On essentially supernilpotent rings and their dual,* Comm. Algebra **23** (1995), 4473–4479.

E. R. Puczyłowski, Institute of Mathematics University of Warsaw, P.O. Box 127 02-097 Warszawa, Poland
E-mail address: edmundp@mimuw.edu.pl

R. Wiegandt, A. Rényi Institute of Mathematics, Hungarian Academy of Sciences Banacha 2, 1364 Budapest, Hungary
E-mail address: wiegandt@renyi.hu

Publications of Kostia Beidar

Since 1977 Kostia Beidar published over 130 papers in more than 20 journals. His monograph *"Rings with generalized identities"* (joint with W. S. Martindale and A. V. Mikhalev) published in 1996 became the main tool for researchers working in this area. It is not surprising that the book was quoted more than 120 times for less than 10 years!

Kostia cooperated with mathematicians from many countries: Austria, Canada, China, Germany, Hong Kong, Hungary, India, Moldova, New Zealand, Poland, Russia, Slovakia, Slovenia, Spain, Sweden, Taiwan, UK and USA. Most of his coauthors became his good friends.

1. Monographs

[1] Beidar, K. I.; Latyshev, V. N.; Markov, V. T.; Mikhalev, A. V.; Skornyakov, L. A.; Tuganbaev, A. A. Associative rings. (Russian) Algebra. Topology. Geometry, Vol. 22, 3–115, 267, Itogi Nauki i Tekhniki, Akad. Nauk SSSR, Vsesoyuz. Inst. Nauchn. i Tekhn. Inform., Moscow, 1984.

[2] Beidar, K. I.; Martindale, W. S., 3rd; Mikhalev, A. V. Rings with generalized identities. Monographs and Textbooks in Pure and Applied Mathematics, 196. Marcel Dekker, Inc., New York, 1996. xiv+522 pp. ISBN: 0-8247-9325-0.

2. Generalized identities

[3] Beidar, K. I.; Ten, V. D. *On the local finiteness of some PI-algebras.* (Russian) Sibirsk. Mat. Ž. **18** (1977), no. 4, 934–938, 959.

[4] Beidar, K. I. *Rings with generalized identities. I.* (Russian) Vestnik Moskov. Univ. Ser. I Mat. Meh. 1977, no. 2, 19–26.

[5] Beidar, K. I. *Rings with generalized identities. II.* (Russian) Vestnik Moskov. Univ. Ser. I Mat. Meh. 1977, no. 3, 30–37.

[6] Beidar, K. I. *Semiprime rings with a generalized identity.* (Russian) Uspehi Mat. Nauk **32** (1977), no. 4(196), 249–250.

[7] Beidar, K. I. *Rings with generalized identities. III.* (Russian) Vestnik Moskov. Univ. Ser. I Mat. Mekh. 1978, no. 4, 66–73.

[8] Beidar, K. I. *Quotient rings of semiprime rings.* (Russian) Vestnik Moskov. Univ. Ser. I Mat. Mekh. 1978, no. 5, 36–43.

[9] Beidar, K. I. *Classical rings of quotients of PI-algebras.* (Russian) Uspekhi Mat. Nauk **33** (1978), no. 6(204), 197–198.

[10] Beidar, K. I.; Mikhalev, A. V.; Salavova, K. *Generalized identities and semiprime rings with involution.* (Russian) Uspekhi Mat. Nauk **35** (1980), no. 1(211), 222.

[11] Beidar, K. I. *Rings with generalized identities. IV.* (Russian) Vestnik Moskov. Univ. Ser. I Mat. Mekh. 1980, no. 4, 3–6, 98.

[12] Beidar, K. I.; Mikhalev, A. V.; Salavova, K. *Generalized identities and semiprime rings with involution.* Math. Z. **178** (1981), no. 1, 37–62.

[13] Beidar, K. I. *Idempotents in rings with a polynomial identity.* (Russian) Comment. Math. Univ. Carolin. **22** (1981), no. 4, 755–759.

[14] Beidar, K. I.; Mikhalev, A. V. *Semiprime rings with bounded indexes of nilpotent elements.* (Russian) Trudy Sem. Petrovsk. No. 13 (1988), 237–249, 259–260; translation in J. Soviet Math. **50** (1990), no. 2, 1518–1526.

[15] Beidar, K. I. *Classical quotient rings of semiprime PI-rings.* (Russian) Algebra i Logika **32** (1993), no. 1, 3–16, 111; translation in Algebra and Logic 32 (1993), no. 1, 1–8.

[16] Beidar, K. I.; Mikhalev, A. V. *Generalized polynomial identities and rings that are sums of two subrings.* (Russian) Algebra i Logika **34** (1995), no. 1, 3–11, 118; translation in Algebra and Logic 34 (1995), no. 1, 1–5.

[17] Beidar, K. I.; Fong, Y.; Knauer, U.; Mikhalev, A. V.; Shum, K. P. *Semigroups with generalized polynomial identities.* First International Tainan-Moscow Algebra Workshop (Tainan, 1994), 147–157, de Gruyter, Berlin, 1996.

[18] Beidar, K. I.; Fong, Y.; Bokut, L. A. *Prime rings with semigroup generalized identity.* Comm. Algebra **28** (2000), no. 3, 1497–1501.

[19] Beidar, K. I.; Ke, W.-F.; Liu, C.-H. *On nil subsemigroups of rings with group identities.* Comm. Algebra **30** (2002), no. 1, 347–352.

[20] Beidar, K. I.; Chen, T.-S.; Fong, Y.; Ke, W.-F. *On graded polynomial identities with an anti-automorphism.* J. Algebra **256** (2002), no. 2, 542–555.

[21] Beidar, K. I.; Chebotar, M. A. *When is a graded PI algebra a PI algebra?* Comm. Algebra **31** (2003), no. 6, 2951–2964.

3. Actions on rings. Hopf algebras

[22] Beidar, K. I. *The ring of invariants under the action of a finite group of automorphisms of a ring.* (Russian) Uspehi Mat. Nauk **32** (1977), no. 1(193), 159–160.

[23] Beidar, K. I. *Associative rings and finite groups of automorphisms.* (Russian) Trudy Sem. Petrovsk. No. 4 (1978), 33–44.

[24] Beidar, K. I.; Stolin, A. A. *The Picard group of a projective limit of rings.* (Russian) Operators in function spaces and problems in function theory (Russian), 126–131, 148, "Naukova Dumka", Kiev, 1987.

[25] Beidar, K. I.; Grzeszczuk, P. *Actions of Lie algebras on rings without nilpotent elements.* Algebra Colloq. **2** (1995), no. 2, 105–116.

[26] Beidar, K. I.; Fong, Y.; Stolin, A. *On Frobenius algebras and the quantum Yang-Baxter equation.* Trans. Amer. Math. Soc. **349** (1997), no. 9, 3823–3836.

[27] Beidar, K. I.; Fong, Y.; Stolin, A. A. *On antipodes and integrals in Hopf algebras over rings and the quantum Yang-Baxter equation.* J. Algebra **194** (1997), no. 1, 36–52.

[28] Beidar, K.; Fong, Y.; Stolin, A. *Symmetric algebras and Yang-Baxter equation*. The Proceedings of the 16th Winter School "Geometry and Physics" (Srn 1996). Rend. Circ. Mat. Palermo (2) Suppl. No. 46 (1997), 15–28.

[29] Beidar, K. I.; Torrecillas, B. *On actions of Hopf algebras with cocommutative coradical*. J. Pure Appl. Algebra **161** (2001), no. 1-2, 13–30.

4. Module theory

[30] Beidar, K. I. *Modules over commutative semiprime rings*. (Russian) Mat. Zametki **29** (1981), no. 1, 15–18, 154.

[31] Beidar, K. I.; Salavova, K. *A class of rings with an essential right socle*. (Russian) Acta Math. Hungar. **53** (1989), no. 1-2, 55–59.

[32] Beidar, K. I.; Wisbauer, R. *Strictly semiprime modules and rings*. (Russian) Uspekhi Mat. Nauk **48** (1993), no. 1(289), 161–162; translation in Russian Math. Surveys 48 (1993), no. 1, 163–164.

[33] Beidar, K.; Wisbauer, R. *Strongly and properly semiprime modules and rings*. Ring theory (Granville, OH, 1992), 58–94, World Sci. Publishing, River Edge, NJ, 1993.

[34] Beidar, K. I.; Knauer, U.; Mikhalev, A. V. *An example of two von Neumann regular rings with nonisomorphic Morita equivalent multiplicative semigroups*. Semigroup Forum **48** (1994), no. 3, 381–383.

[35] Beidar, K. I.; Wisbauer, R. *Properly semiprime self-pp-modules*. Comm. Algebra **23** (1995), no. 3, 841–861.

[36] Beidar, K. I.; Mikhalev, A. V. *Anti-isomorphisms of endomorphism rings of modules and the Morita anti-equivalence*. (Russian) Uspekhi Mat. Nauk **50** (1995), no. 1(301), 187–188; translation in Russian Math. Surveys 50 (1995), no. 1, 191–192.

[37] Beidar, K. I.; Mikhalev, A. V. *Anti-isomorphisms, induced by Morita anti-equivalences, of endomorphism rings of modules that are close to free*. (Russian) Tr. Semin. im. I. G. Petrovskogo No. 19 (1996), 338–344, 349; translation in J. Math. Sci. (New York) **85** (1997), no. 6, 2450–2453.

[38] Beidar, K. I.; Ke, W.-F. *On essential extensions of direct sums of injective modules*. Arch. Math. (Basel) **78** (2002), no. 2, 120–123.

[39] Beidar, K. I.; Fong, Y.; Ke, W.-F.; Jain, S. K. *An example of a right q-ring*. Israel J. Math. **127** (2002), 303–316.

[40] Beidar, K. I.; Jain, S. K.; Kanwar, P.; Srivastava, J. B. *CS matrix rings over local rings*. J. Algebra **264** (2003), no. 1, 251–261.

[41] Beidar, K. I.; Jain, S. K. *The structure of right continuous right π-rings*. Comm. Algebra **32** (2004), no. 1, 315–332.

[42] Beidar, K. I.; Jain, S. K.; Kanwar, P.; Srivastava, J. B. *Semilocal CS matrix rings of order > 1 over group algebras of solvable groups are selfinjective*. J. Algebra **275** (2004), no. 2, 856–858.

[43] Beidar, K. I.; Jain, S. K.; Kanwar, P. *Nonsingular CS-rings coincide with tight PP rings*. J. Algebra **282** (2004), no. 2, 626–637.

[44] Beidar, K. I.; Jain, S. K. *When is every module with essential socle a direct sums of quasi-injectives?* Comm. Algebra **33** (2005), no. 11, 4251–4258.

[45] Beidar, K. I.; Maŕki, L.; Mlitz, R.; Wiegandt, R. *Primitive involution rings.* Acta Math. Hungar. **109** (2005), no. 4, 357–368.

[46] Beidar, K. I.; Jain, S. K.; Srivastava, J. B. *Essential extensions of a direct sum of simple modules.* Contemporary Math. To appear.

5. Radical theory

[47] Beidar, K. I. *Radicals of finitely generated algebras.* (Russian) Uspekhi Mat. Nauk **36** (1981), no. 6(222), 203–204.

[48] Beidar, K. I. *A chain of Kurosh may have an arbitrary finite length.* Czechoslovak Math. J. **32(107)** (1982), no. 3, 418–422.

[49] Beidar, K. I.; Salavova, K. *Some examples of supernilpotent nonspecial radicals.* Acta Math. Acad. Sci. Hungar. **40** (1982), no. 1-2, 109–112.

[50] Beidar, K. I.; Salavova, K. *The lattices of N-radicals, left strong radicals, and left hereditary radicals.* (Russian) Acta Math. Hungar. **42** (1983), no. 1-2, 81–95.

[51] Beidar, K. I. *Atoms in the "lattice" of radicals.* (Russian) Mat. Issled. No. 85, Algebry, Koltsa i Topologii (1985), 21–31, 152.

[52] Beidar, K. I. *Examples of rings and radicals.* Radical theory (Eger, 1982), 19–46, Colloq. Math. Soc. Janos Bolyai, 38, North-Holland, Amsterdam, 1985.

[53] Beidar, K. I. *Semisimple classes of algebras and a lower radical.* (Russian) Mat. Issled. No. 105, Moduli, Algebry, Topol. (1988), 13–29, 194.

[54] Beidar, K. I. *The intersection property for radicals.* (Russian) Uspekhi Mat. Nauk **44** (1989), no. 1(265), 187–188; translation in Russian Math. Surveys 44 (1989), no. 1, 235–236.

[55] Beidar, K. I.; Glavatskii, S. T.; Mikhalev, A. V. *Semisimple classes and lower radicals of topological nonassociative algebras.* (Russian) Trudy Sem. Petrovsk. No. 14 (1989), 250–261, 268; translation in J. Soviet Math. 51 (1990), no. 4, 2487–2496.

[56] Beidar, K. I.; Salavova, K. *Additive radicals.* Czechoslovak Math. J. **39(114)** (1989), no. 4, 659–673.

[57] Beidar, K. I.; Trokanova-Salavova, K. *On nil rings satisfying minimum condition on principal right ideals.* Acta Math. Hungar. **55** (1990), no. 3-4, 197–200.

[58] Arnautov, V. I.; Beidar, C. I.; Glavatskii, S. T.; Mikhalev, A. V. *Intersection property in the radical theory of topological algebras.* Proceedings of the International Conference on Algebra, Part 2 (Novosibirsk, 1989), 205–225, Contemp. Math., 131, Part 2, Amer. Math. Soc., Providence, RI, 1992.

[59] Arnautov, V. I.; Beidar, K. I.; Glavatskii, S. T.; Mikhalev, A. V. *The intersection property in the theory of radicals of topological algebras.* J. Soviet Math. **60** (1992), no. 6, 1782–1789.

[60] Beidar, K. I. *On essential extensions, maximal essential extensions and iterated maximal essential extensions in radical theory.* Theory of radicals (Szekszard, 1991), 17–26, Colloq. Math. Soc. Janos Bolyai, 61, North-Holland, Amsterdam, 1993.

[61] Beidar, K. I. *On questions of B. J. Gardner and A. D. Sands.* J. Austral. Math. Soc. Ser. A **56** (1994), no. 3, 314–319.

[62] Beidar, K. I.; Mikhalev, A. V. *On topologization of rings with infinite lower Baer radical.* Second International Conference on Algebra (Barnaul, 1991), 43–47, Contemp. Math., 184, Amer. Math. Soc., Providence, RI, 1995.

[63] Beidar, K. I.; Fong, Y.; Wang, C. S. *On the lattice of strong radicals.* J. Algebra **180** (1996), no. 2, 334–340.

[64] Beidar, K. I.; Wiegandt, R. *Radicals induced by the total of rings.* Beitrage Algebra Geom. **38** (1997), no. 1, 149–159.

[65] Beidar, K. I.; Fong, Y.; Ke, W.-F. *On complemented radicals.* J. Algebra **201** (1998), no. 1, 328–356.

[66] Beidar, K. I.; Fong, Y. *On radicals of monomial algebras.* Comm. Algebra **26** (1998), no. 12, 3913–3919.

[67] Beidar, K. I. *On principally hereditary radicals.* Comm. Algebra **26** (1998), no. 12, 3899–3912.

[68] Beidar, K. I.; Wiegandt, R. *Radical assignments and radical classes.* Bul. Acad. Ştiinţe Repub. Mold. Mat. 1999, no. 2, 17–27, 113, 116.

[69] Beidar, K. I.; Fong, Y.; Puczyłowski, E. R. *Polynomial rings over nil rings cannot be homomorphically mapped onto rings with nonzero idempotents.* J. Algebra **238** (2001), no. 1, 389–399.

[70] Beidar, K. I.; Puczyłowski, E. R.; Wiegandt, R. *Radicals and polynomial rings.* J. Aust. Math. Soc. **72** (2002), no. 1, 23–31.

[71] Beidar, K. I.; Ke, W.-F.; Puczyłowski, E. R. *On subhereditary radicals and reduced rings.* Proc. Roy. Soc. Edinburgh Sect. A **132** (2002), no. 2, 255–266.

[72] Beidar, K. I.; Fong, Y.; Puczyłowski, E. R. *On essential extensions of reduced rings and domains.* Arch. Math. (Basel) **83** (2004), no. 4, 344–352.

[73] Beidar, K. I.; Ke, W.-F.; Puczyłowski, E. R. *On matrix rings and subhereditary radicals.* Comm. Algebra **32** (2004), no. 7, 2827–2839.

6. Logical aspects of ring theory

[74] Beidar, K. I.; Mikhalev, A. V. *Orthogonal completeness and minimal prime ideals.* (Russian) Trudy Sem. Petrovsk. No. 10 (1984), 227–234, 239.

[75] Beidar, K. I.; Mikhalev, A. V. *Orthogonal completeness and algebraic systems.* (Russian) Uspekhi Mat. Nauk **40** (1985), no. 6(246), 79–115, 199.

[76] Beidar, K. I.; Mikhalev, A. V. *Uniform boundedness almost everywhere for orthogonally complete algebraic systems.* (Ukrainian) Vīsnik Kiev. Unīv. Ser. Mat. Mekh. No. 27 (1985), 15–17, 123.

[77] Beidar, K. I.; Mikhalev, A. V. *The method of orthogonal completeness in the structure theory of rings.* Algebra. 1. J. Math. Sci. **73** (1995), no. 1, 1–46.

[78] Beidar, K. I.; Mikhalev, A. V.; Puninskii, G. E. *Logical aspects of the theory of rings and modules.* (Russian) Fundam. Prikl. Mat. **1** (1995), no. 1, 1–62.

7. Nonassociative algebras

[79] Beidar, K. I. *On A. I. Mal'tsev's theorems on matrix representations of algebras*. (Russian) Uspekhi Mat. Nauk **41** (1986), no. 5(251), 161–162.

[80] Beidar, K. I.; Mikhalev, A. V.; Slin'ko, A. M. *A primality criterion for nondegenerate alternative and Jordan algebras*. (Russian) Trudy Moskov. Mat. Obshch. **50** (1987), 130–137, 261; translation in Trans. Moscow Math. Soc. 1988, 129–137.

[81] Beidar, K. I.; Mikhalev, A. V. *The structure of nondegenerate alternative algebras*. (Russian) Trudy Sem. Petrovsk. No. 12 (1987), 59–74, 243; translation in J. Soviet Math. 47 (1989), no. 3, 2525–2536.

[82] Beidar, K. I. *The Andrunakievich lemma and Jordan algebras*. (Russian) Uspekhi Mat. Nauk **45** (1990), no. 4(274), 137–138; translation in Russian Math. Surveys 45 (1990), no. 4, 159.

[83] Beidar, K. I.; Wiegandt, R. *Splitting theorems for nonassociative rings*. Publ. Math. Debrecen **38** (1991), no. 1-2, 121–143.

[84] Beidar, K. I. *Classical localizations of alternative algebras*. J. Math. Sci. **69** (1994), no. 3, 1098–1104.

[85] Beidar, K. I.; Martindale, W. S., 3rd; Mikhalev, A. V. *Lie isomorphisms in prime rings with involution*. J. Algebra **169** (1994), no. 1, 304–327.

[86] Beidar, K. I.; Pikhtil'kov, S. A. *On the prime radical of special Lie algebras*. (Russian) Uspekhi Mat. Nauk **49** (1994), no. 1(295), 233; translation in Russian Math. Surveys 49 (1994), no. 1, 225–226.

[87] Beidar, K. I.; Chebotar, M. A. *On Lie-admissible algebras whose commutator Lie algebras are Lie subalgebras of prime associative algebras*. J. Algebra **233** (2000), no. 2, 675–703.

[88] Beidar, K. I.; Brešar, M.; Chebotar, M. A. *Jordan isomorphisms of triangular matrix algebras over a connected commutative ring*. Linear Algebra Appl. **312** (2000), no. 1-3, 197–201.

[89] Beidar, K. I.; Pikhtil'kov, S. A. *The prime radical of special Lie algebras*. (Russian) Fundam. Prikl. Mat. **6** (2000), no. 3, 643–648.

[90] Beidar, K. I.; Chebotar, M. A. *On surjective Lie homomorphisms onto Lie ideals of prime rings*. Comm. Algebra **29** (2001), no. 10, 4775–4793.

[91] Beidar, K. I.; Chebotar, M. A. *On Lie derivations of Lie ideals of prime algebras*. Israel J. Math. **123** (2001), 131–148.

[92] Beidar, K. I.; Brešar, M.; Chebotar, M. A.; Martindale, W. S., 3rd. *On Herstein's Lie map conjectures. I*. Trans. Amer. Math. Soc. **353** (2001), no. 10, 4235–4260.

[93] Beidar, K. I.; Brešar, M.; Chebotar, M. A.; Martindale, W. S., 3rd. *On Herstein's Lie map conjectures. II*. J. Algebra **238** (2001), no. 1, 239–264.

[94] Beidar, K. I.; Brešar, M.; Chebotar, M. A.; Martindale, W. S., 3rd. *On Herstein's Lie map conjectures. III*. J. Algebra **249** (2002), no. 1, 59–94.

[95] Beidar, K. I.; Zaĭtsev, M. V.; Pikhtil'kov, S. A. *Lie algebras with the maximality condition for abelian subalgebras*. (Russian) Vestnik Moskov. Univ. Ser. I Mat. Mekh. 2002, , no. 5, 27–32, 72; translation in Moscow Univ. Math. Bull. **57** (2002), no. 5, 29–33 (2003).

[96] Beidar, K. I.; Brešar, M.; Chebotar, M. A. *Jordan superhomomorphisms*. Comm. Algebra **31** (2003), no. 2, 633–644.

[97] Beidar, K. I.; Brešar, M.; Chebotar, M. A.; Martindale, W. S., 3rd. *Polynomial preserving maps on certain Jordan algebras*. Israel J. Math. 141 (2004), 285–313.

[98] Beidar, K. I.; Chebotar, M. A.; Fong, Y.; Ke, W.-F. *On some Lie-admissible subalgebras of matrix algebras.* (Russian) Sovrem. Mat. Prilozh. No. 13, Algebra (2004), 71–78; translation in J. Math. Sci. (N. Y.) **131** (2005), no. 5, 5939–5947.

8. General ring theory

[99] Beidar, K. I.; Mikhalev, A. V. *Additive continuation of derivations of multiplicative semi-groups.* (Russian) Mechanics and applied mathematics (Russian), 112–114, Priok. Knizhn. Izdat., Tula, 1988.

[100] Beidar, K. I.; Glavatskii, S. T.; Mikhalev, A. V. *Varieties of topological Ω-groups.* (Russian) Izv. Vyssh. Uchebn. Zaved. Mat. 1989, , no. 6, 40–42; translation in Soviet Math. (Iz. VUZ) **33** (1989), no. 6, 35–37.

[101] Beidar, K. I.; Mikhalev, A. V. *Generalized Goldie rings. I, II.* (Russian) Abelian groups and modules, No. 8 (Russian), 17–34, 35–53, 172, Tomsk. Gos. Univ., Tomsk, 1989.

[102] Beidar, C. I.; Mikhalev, A. V. *On Mal'cev's theorem on elementary equivalence of linear groups.* Proceedings of the International Conference on Algebra, Part 1 (Novosibirsk, 1989), 29–35, Contemp. Math., 131, Part 1, Amer. Math. Soc., Providence, RI, 1992.

[103] Beidar, K. I.; Wiegandt, R. *Rings with involution and chain conditions.* J. Pure Appl. Algebra **87** (1993), no. 3, 205–220.

[104] Beidar, K. I.; Wiegandt, R. *Rings with involution and conditions for bi-ideal chains.* (Russian) Uspekhi Mat. Nauk **48** (1993), no. 5(293), 159–160; translation in Russian Math. Surveys 48 (1993), no. 5, 161–162.

[105] Beidar, K. I.; Markov, V. T. *A semiprime PI ring which has a faithful module with Krull dimension is a Goldie ring.* (Russian) Uspekhi Mat. Nauk **48** (1993), no. 6(294), 141–142; translation in Russian Math. Surveys 48 (1993), no. 6, 158.

[106] Beidar, K. I.; Puczyłowski, E. R.; Smith, P. F. *Krull dimension of modules over involution rings.* Proc. Amer. Math. Soc. **121** (1994), no. 2, 391–397.

[107] Beidar, K. I.; Puczyłowski, E. R.; Smith, P. F. *Krull dimension of modules over involution rings. II.* Proc. Amer. Math. Soc. **125** (1997), no. 2, 355–361.

[108] Beidar, K. I. *On rings with zero total.* Beitrage Algebra Geom. **38** (1997), no. 2, 233–239.

[109] Beidar, K. I.; Brešar, M. *Applying the density theorem for derivations to range inclusion problems.* Studia Math. **138** (2000), no. 1, 93–100.

[110] Beidar, K. I.; Brešar, M. *Extended Jacobson density theorem for rings with derivations and automorphisms.* Israel J. Math. **122** (2001), 317–346.

[111] Beidar, K. I.; Brešar, M.; Fong, Y. *Extended Jacobson density theorem for Lie ideals of rings with automorphisms.* Publ. Math. Debrecen **58** (2001), no. 3, 325–335.

[112] Beidar, K. I.; Kasch, F. *Good conditions for the total.* International Symposium on Ring Theory (Kyongju, 1999), 43–65, Trends Math., Birkhauser Boston, Boston, MA, 2001.

[113] Beidar, K. I.; Wisbauer, R. *On uniform bounds of primeness in matrix rings.* J. Aust. Math. Soc. **76** (2004), no. 2, 167–174.

[114] Beidar, K. I.; O'Meara, K. C.; Raphael, R. M. *On uniform diagonalisation of matrices over regular rings and one-accessible regular algebras.* Comm. Algebra **32** (2004), no. 9, 3543–3562.

[115] Beidar, K. I.; Marki, L.; Mlitz, R.; Wiegandt, R. *Primitive involution rings.* Acta Math. Hungar. **109** (2005), no. 4, 357–368.

9. Near-rings

[116] Beidar, K. I.; Fong, Y.; Shum, K. P. *On the hearts of subdirectly irreducible near-rings.* Southeast Asian Bull. Math. **18** (1994), no. 2, 5–9.

[117] Beidar, K. I.; Fong, Y.; Ke, W.-F.; Liang, S.-Y. *Nearring multiplications on groups.* Comm. Algebra **23** (1995), no. 3, 999–1015.

[118] Beidar, K. I.; Fong, Y.; Wang, X. K. *Posner and Herstein theorems for derivations of 3-prime near-rings.* Comm. Algebra **24** (1996), no. 5, 1581–1589.

[119] Beidar, K.; Fong, Y.; Ke, W.-F. *On finite circular planar nearrings.* J. Algebra **185** (1996), no. 3, 688–709.

[120] Beidar, K. I.; Fong, Y.; Ke, W.-F. *On the simplicity of centralizer nearrings.* First International Tainan-Moscow Algebra Workshop (Tainan, 1994), 139–146, de Gruyter, Berlin, 1996.

[121] Beidar, K. I.; Fong, Y.; Ke, W.-F.; Wu, W.-R. *On semi-endomorphisms of groups.* Comm. Algebra **27** (1999), no. 5, 2193–2205.

[122] Beidar, K. I.; Fong, Y.; Ke, W.-F. *Maximal right nearring of quotients and semigroup generalized polynomial identity.* Results Math. **42** (2002), no. 1-2, 12–27.

[123] Beidar, K. I.; Ke, W.-F.; Liu, C.-H.; Wu, W.-R. *Automorphism groups of certain simple 2-$(q, 3, \lambda)$ designs constructed from finite fields.* Finite Fields Appl. **9** (2003), no. 4, 401–412.

[124] Beidar, K. I.; Ke, W.-F.; Kiechle, H. *Circularity of finite groups without fixed points.* Monatsh. Math. **144** (2005), no. 4, 265–273.

[125] Beidar, K. I.; Ke, W.-F.; Kiechle, H. *Automorphisms of certain design groups II.* J. Algebra. To appear.

10. Functional identities

[126] Beidar, K. I.; Fong, Y.; Lee, P.-H.; Wong, T.-L. *On additive maps of prime rings satisfying the Engel condition.* Comm. Algebra **25** (1997), no. 12, 3889–3902.

[127] Beidar, K. I. *On functional identities and commuting additive mappings.* Comm. Algebra **26** (1998), no. 6, 1819–1850.

[128] Beidar, K. I.; Martindale, W. S., III. *On functional identities in prime rings with involution.* J. Algebra **203** (1998), no. 2, 491–532.

[129] Beidar, K. I.; Brešar, M.; Chebotar, M. A. *Generalized functional identities with (anti-)automorphisms and derivations on prime rings. I.* J. Algebra **215** (1999), no. 2, 644–665.

[130] Beidar, K. I.; Fong, Y. *On additive isomorphisms of prime rings preserving polynomials.* J. Algebra **217** (1999), no. 2, 650–667.

[131] Beidar, K. I.; Brešar, M.; Chebotar, M. A.; Martindale, W. S., III. *On functional identities in prime rings with involution. II.* Comm. Algebra **28** (2000), no. 7, 3169–3183.

[132] Beidar, K. I.; Chang, S.-C.; Chebotar, M. A.; Fong, Y. *On functional identities in left ideals of prime rings.* Comm. Algebra **28** (2000), no. 6, 3041–3058.

[133] Beidar, K. I.; Brešar, M.; Chebotar, M. A. *Functional identities on upper triangular matrix algebras*. Algebra, 15. J. Math. Sci. (New York) **102** (2000), no. 6, 4557–4565.

[134] Beidar, K. I.; Chebotar, M. A. *On functional identities and d-free subsets of rings. I, II*. Comm. Algebra **28** (2000), no. 8, 3925–3951, 3953–3972.

[135] Beidar, K. I.; Brešar, M.; Chebotar, M. A. *Functional identities revised: the fractional and the strong degree*. Comm. Algebra **30** (2002), no. 2, 935–969.

[136] Beidar, K. I.; Brešar, M.; Chebotar, M. A.; Fong, Y. *Applying functional identities to some linear preserver problems*. Pacific J. Math. **204** (2002), no. 2, 257–271.

[137] Beidar, K. I.; Brešar, M.; Chebotar, M. A. *Functional identities with r-independent coefficients*. Comm. Algebra **30** (2002), no. 12, 5725–5755.

[138] Beidar, K. I.; Lin, Y.-F. *On surjective linear maps preserving commutativity*. Proc. Roy. Soc. Edinburgh Sect. A **134** (2004), no. 6, 1023–1040.

[139] Beidar, K. I.; Mikhalev, A. V.; Chebotar, M. A. *Functional identities in rings and their applications*. (Russian) Uspekhi Mat. Nauk **59** (2004), no. 3(357), 3–30; translation in Russian Math. Surveys 59 (2004), no. 3, 403–428.

[140] Beidar, K. I.; Lin, Y.-F. *Maps characterized by action on Lie zero products*. Comm. Algebra **33** (2005), no. 8, 2697–2703.